8th International Conference on Intelligent Textiles & Mass Customisation 2022

IOP Conference Series: Materials Science and Engineering
Volume 1266

Montreal, Canada
18 – 21 September 2022

ISBN: 978-1-7138-6765-4
ISSN: 1757-8981

Printed from e-media with permission by:

Curran Associates, Inc.
57 Morehouse Lane
Red Hook, NY 12571

Some format issues inherent in the e-media version may also appear in this print version.

Institute of Physics
Dirac House, Temple Back
Bristol BS1 6BE UK

Phone: 44 1 17 929 7481
Fax: 44 1 17 920 0979

techtracking@iop.org

Additional copies of this publication are available from:

Curran Associates, Inc.
57 Morehouse Lane
Red Hook, NY 12571 USA
Phone: 845-758-0400
Fax: 845-758-2633
Email: curran@proceedings.com
Web: www.proceedings.com

TABLE OF CONTENTS

IOP Conf. Series: Materials Science and Engineering 1266 (2023) 011001 doi:10.1088/1757-899X/1266/1/011001

2022 ITMC
Conference & Smart Textiles Salon

PREFACE

The 8th Edition of International Conference on Intelligent Textiles and Mass Customization (ITMC 2022), which was held at the beautiful city of Montreal (Canada), from the 21st to the 23rd of September 2021, targeted guests from various branches and disciplines related to the textile industry. Its interdisciplinary approach is the key to maximizing the potential and development of textile materials and tools for various applications. The purpose of the conference is to explore new ideas, effective solutions and collaborative partnerships for business growth by catalyzing the creation of a beneficial synergy between designers, manufacturers, suppliers and end users of all sectors and making full use of this potential.

The International Conference on Intelligent Textiles and Mass Customization (ITMC) meets every two years, where the organization rotates among the 5 coordinating countries (Belgium, Canada, France, Japan and Morocco). On behalf of the Conference organizers, we are honored to invite all interested business, research institutions and organizations from around the globe to participate in a lively exchange of ideas and experiences featured at the ITMC2022 Conference.

ITMC conference themes are axed on intelligent textiles and mass customization. During two days, Inspiring speakers from industries, academies, governments and societies have shined the light over new chances and challenges, bringing global statistics and success stories about cutting edge science and technology. The innovation brought to the table of discussion will bloom through cooperation, policy, education and training and rise via an outstanding interaction between speakers and participants, guaranteed through Novel IT tools. Participants were also invited to show their prototypes during the Smart Textiles Salon.

IOP Conf. Series: Materials Science and Engineering 1266 (2023) 011002 doi:10.1088/1757-899X/1266/1/011002

Peer Review Statement

All papers published in this volume have been reviewed through processes administered by the Editors. Reviews were conducted by expert referees to the professional and scientific standards expected of a proceedings journal published by IOP Publishing Publishing.

- **Type of peer review:** Single Anonymous
- **Conference submission management system:** In-house process
- **Number of submissions received:** 28
- **Number of submissions sent for review:** 28
- **Number of submissions accepted:** 26
- **Acceptance Rate (Submissions Accepted / Submissions Received × 100):** 92.9
- **Average number of reviews per paper:** 2.6
- **Total number of reviewers involved:** 13
- **Contact person for queries:**
  Name: Vincent Deregnaucourt
  Email: vderegnaucourt@gcttg.com
  Affiliation: CTT Group

IOP Conf. Series: Materials Science and Engineering 1266 (2023) 012001 doi:10.1088/1757-899X/1266/1/012001

3D Printing of Conductive Flexible Filaments for E-Textile Applications

Gozde Goncu-Berk

Department of Design, University of California-Davis, CA, USA

ggoncuberk@ucdavis.edu

Abstract. Electronic textiles (e-textiles) can incorporate conductive materials at all levels of integration, from fibers to yarns to fabric itself. There are many ways to connect the textile elements to electronic components with interconnect mechanisms from mechanical gripping to welding, to gluing, to printing, to embroidery, knitting, and weaving. 3D printing method offers the possibility of creating flexible and stretchable interconnects for e-textiles applications. This study explored 3D printing of flexible conductive filaments on fabric to create interconnects for hard electrical components as well as transmission lines and switches for electronic textile applications. NinjaTek Eel and Palmiga TPU based conductive filaments were printed on polyester knit fabric. Electrical characterization measurements as well as visual and haptic analysis of printed samples were conducted. The results showed that TPU based flexible conductive filaments offer possibilities of direct 3D printing onto textiles for electronic textile applications.

1. Introduction

Electronic textiles (e-textiles) can incorporate conductive materials at all levels of integration, from fibers to yarns to fabric itself. E-textiles may also include a number of different electronic components, such as sensors, actuators, and energy sources like batteries or solar cells [1].

There are many ways to connect the textile elements to electronic components, from mechanical gripping, to welding, to gluing, to printing, to embroidery, knitting, and weaving [1]. Interconnects to join electronic and textile elements "while maintaining the functionality of electronics and the flexibility and the robustness of textiles" often require compromise [2]. Interconnects have to withstand common stresses such as stretching, bending, moisture, and abrasion textiles and garments are commonly exposed to [3]. For example, even gluing electronic components using conductive epoxy demonstrates reliable conduction, this type of interconnect is inflexible in nature and bending stress from movement of textiles cause it to crack [4]. The traditional method of soldering only works to create connections on conductive textiles that are made from metal fibers as high temperatures melt mixed fibers. Mechanical grip connections like metal snaps hold the advantage of better conductivity compared to adhesive and conductive thread interconnects [5], however, some studies reported negative impact of snap interconnects on electrical signal quality [4]. Explorations of embroidery to join textiles with electronics yielded flexible and stretchable interconnects [6].

Another method that offers the possibility of creating flexible and stretchable interconnects for e-textiles applications is 3D printing on fabric. This study explores 3D printing of flexible conductive filaments on fabric to create interconnects for hard electrical components as well as transmission lines and switches for electronic textile applications.

2. Materials and Methods

The only two commercially available Thermoplastic Urethane (TPU) conductive filaments were selected for the study as they are flexible and allow bending and flexing with the fabric they are printed on. Table 1 displays mechanical characteristics and printing parameters of the filaments used

Table 1. Mechanical properties of 3D printing filaments.

Filament	Shore Hardness	Diameter(mm)	Extruder Temperature	Printing Bed Temperature
NinjaTek Eel	90A	1.75	220°C – 230°C	Up to 45°C
Palmiga-PI-ETPU	95A	1.75	210–230 °C	NA

The filaments were printed in 2x2x10mm rectangle form with Fused Deposition Modelling process using a Raise3D N1 two nozzle 3D printing machine. Next, filaments were printed on a polyester knit textile surface in the same form by clamping the textile on the printing bed and extracting the filaments from the nozzle at an optimum distance and temperature. The rectangles printed on the textile surfaces were created as CAD models, exported as .stl files to the IdeaMaker software for slicing into layers. Printed samples were tested for their electrical resistivity in three different conditions, 1. Filament Form, 2. 3D Printed, 3. 3D printed on Polyester Knit Fabric, using GWInstek GDM 8351 multimeter with triple repeat. Samples of 3D printed filaments on polyester knit fabric were evaluated by visual and haptic inspection for each filament's adhesion on the fabric surface and their flexibility. Next, a circuit with raised snap fit interconnects for sewable LEDs and battery holders and a switch were designed, 3D modeled in Rhinoceros software and sliced into layers for 3D printing. The circuit was printed on the polyester knit fabric using both filaments.

3. Results and Discussion

Conductive filaments' resistance values were evaluated under three conditions. For both filaments electrical resistance increased after they were 3D printed compared to their pure filament forms. Resistance also increased more when the filaments were printed on the polyester knitted textile. The increase may be explained by the amount of bonding between the conductive filament and the textile and conductive filament polymer molecules' diffusive penetration into the fabric structure. Therefore, the molecular bonding may affect the resistance of the conductive filament.

Table 2. Resistance of conductive filament.

Sample (Length: 10 mm)	Avg. Resistance (kΩ)
NinjaTek Eel filament	0.75
NinjaTek Eel filament 3D printed	12.8
NinjaTek Eel filament 3D printed on Fabric	18.17
Palmiga filament	2.87
Palmiga filament 3D printed	24.8
Palmiga filament 3D printed on Fabric	34.9

Flexibility of the conductive filaments printed on fabric was evaluated haptically and visually while the fabric hangs under its own weight. A seen in Figure 1, NinjaTek Eel filament demonstrated higher flexibility compared to Palmiga filament and bended more naturally with the fabric. This result was already expected based on shore hardness values of the two filaments where NinaTek demonstrates higher flexibility with the shore hardness of 90A compared to Palmiga's 95A. Based on haptic evaluation, both filaments displayed strong bond to the fabric surface.

Figure 1. Flexion of NinjaTek Eel and Palminga conductive filaments 3D printed on textile surface.

When the circuit design was 3 printed on the fabric surface, all transmission lines were printed successfully with both filament types, however NinjaTek Eel failed in printing the raised snap fit interconnects and the switch (Figure 2). Palmiga filament printed smoothly and displayed a snug connection between the textile surface, the LEDs and the battery holder (Figure 3). The switch was designed with an overhang print configuration with angled inclination of the print wall from the vertical axis. This type of printing is challenging even most commonly used inflexible PLA filaments, however, it was possible to achieve successful result with TPU based conductive Palminga filament. As a result, even though NinjaTek Eel presented lower electrical resistance and better flexibility Palminga proved to be more successful in printing 3D forms.

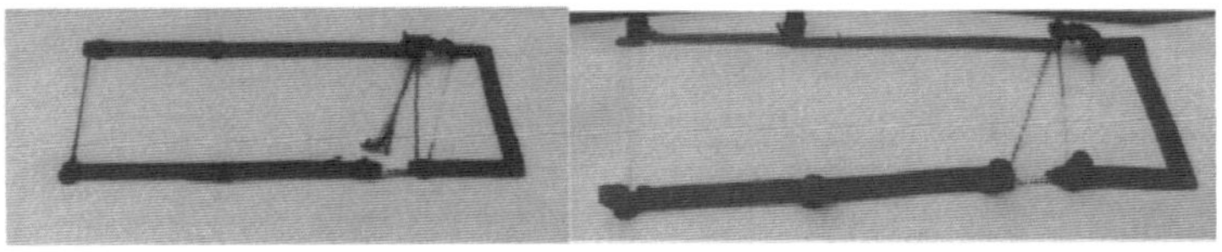

Figure 2. Failed NinjaTek Eel filament print versus successful Palminga filament print.

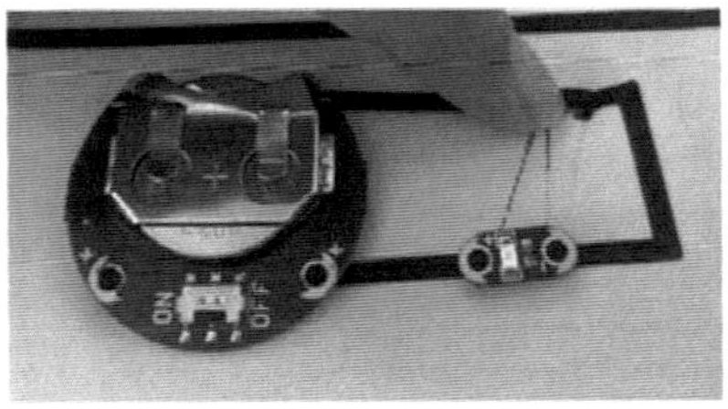

Figure 3. 3D printed circuit with hard components of LED and the battery holder.

4. Conclusion

The results of this study showed that TPU based flexible conductive filaments offer possibilities of direct 3D printing onto textiles for electronic textile applications. Secure temporary interconnects to join hard components to textiles can be achieved which allows ease of laundering by removing the components easily from the textiles. However, research on launderability of textiles with 3D printed structures are needed for future exploration.

References

[1] Castano LM, Flatau AB. Smart fabric sensors and e-textile technologies: a review. Smart Materials and structures. 2014 Apr 1;23(5):053001.

[2] Locher I, Sefar AG. Joining technologies for smart textiles. Multidisciplinary Know-How for Smart-Textiles Developers. 2013 Jan 1:285-305.

[3] Agcayazi T, Chatterjee K, Bozkurt A, Ghosh TK. Flexible interconnects for electronic textiles. Advanced Materials Technologies. 2018 Oct;3(10):1700277.

[4] Pinapati SP, Kaufmann T, Linke I, Ranasinghe D, Fumeaux C. Connection strategies for wearable microwave transmission lines and antennas. In2015 International Symposium on Antennas and Propagation (ISAP) 2015 Nov 9 (pp. 1-4). IEEE.

[5] Posch I, Stark L, Fitzpatrick G. eTextiles: reviewing a practice through its tool/kits. InProceedings of the 23rd international symposium on wearable computers 2019 Sep 9 (pp. 195-205).

[6] Linz T, Vieroth R, Dils C, Koch M, Braun T, Becker KF, Kallmayer C, Hong SM. Embroidered interconnections and encapsulation for electronics in textiles for wearable electronics applications. InAdvances in science and technology 2008 (Vol. 60, pp. 85-94). Trans Tech Publications Ltd.

IOP Conf. Series: Materials Science and Engineering 1266 (2023) 012002 doi:10.1088/1757-899X/1266/1/012002

Embroidered textile electrodes for long-term ECG monitoring

Abreha Bayrau Nigusse [*1,2], Benny Malengier[1], Desalegn Alemu Mengistie [2,3], Ambachew Maru[2], Granch Berhe Tseghail and Lieva Van Langenhove[1]

[1]Department of Materials, Textiles and Chemical Engineering, Ghent University, 9000 Gent, Belgium; (E-mail: abrehabayrau.bayraunigusse@ugent.be, Benny.Malengier@UGent.be, Lieva.VanLangenhove@UGent.be)
[2]Ethiopian Institute of Textile and Fashion Technology, Bahir Dar University, 6000 Bahir Dar, Ethiopia
[3]Materials Engineering Department, California Polytechnic State University, San Luis Obispo, CA 93407, USA; (E-mail: dmengist@calpoly.edu)

Corresponding author's e-mail: AbehaBayrau.BayrauNigusse@UGent.be

Abstract. Electrocardiography (ECG) monitoring is very important for the diagnosis and examination of heart-related diseases. Electrodes are the main components in monitoring and recording ECG signals. In this work, we have developed embroidered electrodes using two types of conductive yarns. ECG detection performance of the electrodes was measured at static and dynamic conditions and results were compared to standard Ag/AgCl electrodes. The electrodes were developed with three different dimensional areas to investigate the effect of the size on ECG acquisition performance. ECG signals collected from both the embroidered and standard Ag/AgCl electrodes have visible P, QRS, and T waveforms. Signals collected using large-size textile electrodes show better signal amplitude, which would reveal that the performance of the electrodes becomes improved with an increase in size. However, signals collect more artifacts dynamic conditions contain motion artifacts, indicating this aspect requires further improvement.

1. Introduction

Continuous electrocardiography (ECG) monitoring is very important for the early detection and treatment of heart arrhythmias. Standard wet electrodes are not suitable for this long-term ECG monitoring because the adhesive gel used in wet Ag/AgCl electrodes gradually dries out during use, affecting signal quality and causing skin irritation [1]. Textile-based electrodes can address these problems [2]. Different types of textile electrodes have been developed and can be found in the literature [3-5].

Weder et al. reported an embroidered ECG electrode with a wetting pad that contains water, in order to moisten the skin continuously during measurement and they reported that moist electrodes provide better signal quality [6]. On the other hand, Kannaian et al. presented embroidered textile electrodes via hand embroidery and studied the effect of yarn consumption and washing on signal quality [7]. They reported that the performance of the electrodes was not severely affected by washing and repeated use.

Overall, textile electrodes are expected to be comfortable, washable, and in good contact with the body skin of the wearer for continuous and constant signal detection (to minimize unwanted signals such as artifacts and noise) and easy to develop. However, at the moment, the development techniques are not easy and the application of textile electrodes for ECG monitoring still needs further investigation before it can be used in clinical practice. As a result, they are not yet routinely used clinically.

This work focuses on the design and development of embroidered textile ECG electrodes. ECG acquisition performance of the developed electrodes was measured at static and dynamic conditions and results were compared to standard Ag/AgCl wet electrodes.

2. Material and Methods

Two types of conductive threads were used as embroidery yarns to develop the electrodes: a fully silver-plated polyamide thread (Maderia HC-40) with a resistance of 136 Ω/m, and an electrically conductive hybrid thread manufactured from ultrafine metallic wires combined with polyamide synthetic filaments from Clevertex®, with a resistance of 9.77 Ω/m. A gray woven fabric having a yarn density of 37 ends/cm and 17 picks/cm has been used as a base material. This gray woven fabric was selected as a base material to minimize the shirking effect and damage to the fabric during machine embroidery.

The electrodes were designed using the Ink/Stitch software, an embroidery plugin for Inkscape, and created using a computerized embroidery machine (Brother PE800). The electrodes were developed in a circular shape with three different dimensional areas (12 cm^2, 8 cm^2, and 4 cm^2 active area) to investigate the effect of the size of the electrodes on ECG detection performance. The electrodes were developed with a stitch length of 1.5 mm.

The ECG detection performance was measured by wrapping the measuring electrodes on the right and left wrist, and the reference electrode on the right leg based on lead I configuration. The electrodes were held in place using an elastic band with Velcro. ECG was obtained during static and dynamic (slow walking) conditions using a PC 80B ECG portable measuring device. For comparison, we also measure ECG using standard Ag/AgCl electrodes. To investigate the effect of washing the textile electrodes were washed according to the standard method ISO 105-C06-A1S, 2010. Each test was recorded for 3 minutes without any skin preparation. The collected signals were evaluated on their quality based on the visibility and amplitude of P, QRS, and T waves. Besides, the heart rate, R-R interval, and signal-to-noise ratio (SNR)of each signal were quantified.

3. Results and discussion

The surface resistance of the electrodes was (0.7 ± 0.) Ω/seq and (0.2 ± 0.04) Ω/seq, respectively for Maderia HC-40 and electroconductive hybrid thread. The results show the embroidered electrodes show high conductivity, which is an important parameter to use as sensors. Figures 1a and b show the ECG signals collected using the embroidered textile electrode of Maderia HC-40 and electroconductive hybrid thread, respectively, and signals from the standard Ag/AgCl electrode are presented in Figure 1c. The signals collected from both the embroidered textile electrodes have clearly visible P, QRS, and T waveforms; they provide comparable signal quality with the standard Ag/AgCl electrodes. There are no missing peaks in both signals collected from the embroidered textiles and standard electrodes. However, signals from the electroconductive hybrid thread were relatively noisy with an SNR of 36.58 dB, compared to signals from Maderia HC-40 electrodes where their SNR was 37.39 dB. As the resistance of the electrodes becomes lower, the ability to collect signals and their stability becomes better [8], from this perspective signals from electroconductive hybrid thread are expected to have better quality due to the lower resistance of the electrodes. But as such yarns contain ultrafine metal micro-wires and textile fibers the signal quality is not as we expect, and relatively the R-peak is not uniform

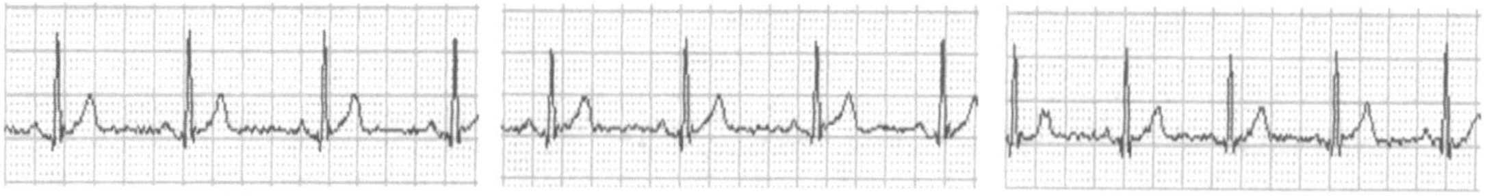

(a) (b) (c)

Figure 1. ECG signal acquired using the: (a) Maderia HC-40 embroidered textile, (b) electroconductive hybrid thread, and (c) Ag/AgCl electrodes

Results in table 1 show a comparison of the ECG signals from the embroidered textile electrodes (Maderia HC-40) having different dimensional areas and the standard Ag/AgCl electrode. Signals collected using large-size textile electrodes (12 cm^2 and 8 cm^2) show better signal amplitude than signals derived from the standard electrode, this could be due to the absence of skin preparation before ECG measurement which is part of the standard clinical procedure for standard electrodes. ECG signals acquired using large-size textile electrodes show higher R-peak amplitude, which would reveal that the performance of the electrodes becomes improved with an increase in size. The amplitude of the R-peak of the signals collected using 12 cm^2 electrodes has no significant difference with signals acquired using 8 cm^2 (p-value = 0.106 at alpha = 0.01), while there was significant difference with signals collected using 4 cm2 area electrodes (p-value < 0.001 at alpha = 0.01). The SNR was 39.31 dB, 37.39 dB, and 37.39 dB for signals acquired using the large, medium, and small size electrodes, respectively which are comparable to signals taken using standard Ag/AgCl electrodes.

	Embroidered electrode size (Area)			
Parameter	**12 cm^2**	**8 cm^2**	**4 cm^2**	**Ag/AgCl**
P-wave (mV)	0.10 ± 0.02	0.11 ± 0.02	0.01 ± 0.02	0.09 ± 0.04
R-peak (mV)	1.32 ± 0.03	1.28 ± 0.03	1.13 ± 0.05	1.28 ± 0.08
T-wave (mV)	0.35 ± 0.01	0.34 ± 0.03	0.29 ± 0.05	0.24 ± 0.01
R-R interval	914.62 ± 45.63	903.13 ± 54.07	914.89 ± 51.2	910.27 ± 45.07
HR (bpm)	76	78	72	77
SNR (dB)	39.31± 3.33	37.39± 1.33	37.18± 4.14	38.46± 2.89

Table 1. Comparison of signals acquired from textile electrodes having different sizes and standard electrodes (values show mean ± SD)

ECG signals collected during walking showed clearly visible major peaks without any missing R-peaks except in some cycles, the small amplitude peaks (P weave) were no longer clearly visible during walking as shown in Figure 2. Still, the textile electrodes acquired comparable signal quality after washing (results not shown here) compared to signals acquired before washing. The results revealed that the developed textile electrodes can be used in static and dynamic conditions.

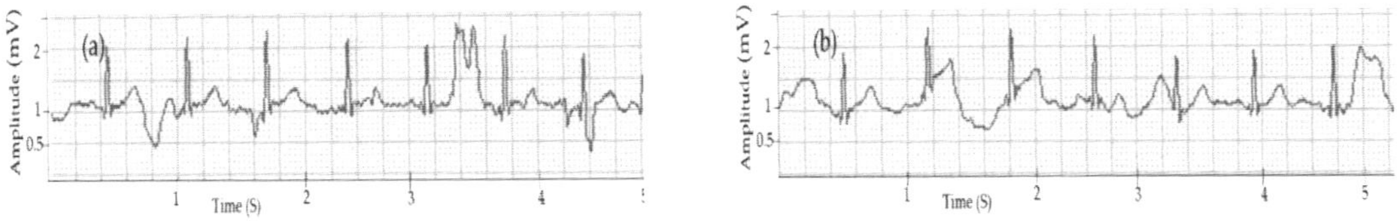

Figure 2. Signals collected while the subject walking: (a) using embroidered textile electrodes (Maderia HC-40); and (b) using Ag/AgCl electrodes

4. Conclusion

In this work, we presented embroidered textile ECG electrodes using silver-plated polyamide (HC40) and electro-conductive hybrid threads. The ECG signals acquired using the textile electrodes have clearly visible P, QRS, and T waveforms without any missing peaks. The amplitudes of the P, QRS, and T waveforms were 0.1, 1.28, and 0.34 mV, respectively, and 37.39 dB SNR for signals from 8 cm^2 area electrodes. Experimental results showed that the proposed electrodes provided comparable signals compared to the standard electrodes. Signals acquired using large-size textile electrodes show better quality. The textile electrodes were successfully used for ECG monitoring at static and dynamic conditions. It should also be possible to use the electrodes for EEG and EMG signals monitoring.

Acknowledgment: We would like to express appreciation for the support of the CoE funded through KFW under grant number No. 51235 in collaboration with the Ethiopian government.

References

[1] Jin-Woo Lee and Kwang-Seok Yun, "ECG Monitoring Garment Using Conductive Carbon Paste for Reduced Motion Artifacts," Polymers, vol. 9, no. 12, p. 439, Sep. 2017.Another reference

[2] C. Lou *et al.*, "Flexible Graphene Electrodes for Prolonged Dynamic ECG Monitoring," *Sensors*, vol. 16, no. 11, p. 1833, Nov. 2016, doi: 10.3390/s16111833.

[3] C. Lou *et al.*, "Flexible Graphene Electrodes for Prolonged Dynamic ECG Monitoring," *Sensors*, vol. 16, no. 11, p. 1833, Nov. 2016, doi: 10.3390/s16111833.

[4] H. Y. Song *et al.*, "Textile electrodes of jacquard woven fabrics for biosignal measurement," *J. Text. Inst.*, vol. 101, no. 8, pp. 758–770, Aug. 2010, doi: 10.1080/00405000903442086.

[5] M. A. Mestrovic, R. J. N. Helmer, L. Kyratzis, and D. Kumar, "Preliminary study of dry knitted fabric electrodes for physiological monitoring," in *2007 3rd International Conference on Intelligent Sensors, Sensor Networks and Information*, Melbourne, Australia, 2007, pp. 601–606, doi: 10.1109/ISSNIP.2007.4496911.

[6] M. Weder *et al.*, "Embroidered Electrode with Silver/Titanium Coating for Long-Term ECG Monitoring," *Sensors*, vol. 15, no. 1, pp. 1750–1759, Jan. 2015, doi: 10.3390/s150101750.

[7] T. Kannaian, R. Neelaveni, and G. Thilagavathi, "Design and development of embroidered textile electrodes for continuous measurement of electrocardiogram signals," *J. Ind. Text.*, vol. 42, no. 3, pp. 303–318, Jan. 2013.

[8] Xiang An, and George K. Stylios, "A hybrid textile electrode for electrocardiogram (ECG) measurement and motion tracking," *Materials*, vol. 11, pp. 1–18, Oct. 2018.

SMART materials & systems: has embracing uncertainty become vital to commercialisation?

Dr M O'Mahony

Textiles Department, Royal College of Art (RCA), Kensington Gore, London SW7 2EU, UK

marie.omahony@rca.ac.uk

Abstract. As humans we have a tendency to seek to confirm data and to avoid disconfirming data. COVID-19 that has thrown plans large and small into disarray, but well before then, the pace of change was already challenging the ability of SMART materials and systems to achieve full scalable commercialisation. While plans may not be feasible, what is possible and indeed necessary is planning. In this presentation I will present a toolkit that takes account of and even embraces uncertainty as a planning method directed at a more effective route for commercialisation and market growth.

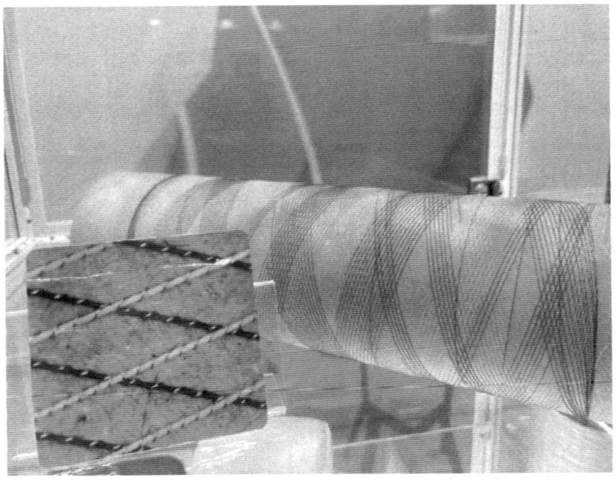

Fig.1. DITF Position-dissolving sensor yarn

1. Introduction
Natural fibers have a long history, so that although undeniably impacted by mechanisation and technology since the First Industrial Revolution, materials such as cotton and wool are inherently 'knowable.' Even as their capabilities are extended, there are inherent qualities, what might be termed a provenance that is singular and so traceable and identifiable. When man-made materials were first developed in the early twentieth century, they looked to mimic natural materials and initially directed at offering low-cost, convenient alternatives to more expensive natural materials such as silk [1]. By the second half of the century their own unique characteristics started to receive attention and their adaptability was seen as offering new design possibilities. This malleability was not without its challenges as Ezio Manzini pointed out in *The Material of Invention* (1989): "a world of nameless

materials is taking shape. These materials create a crisis in the traditional relationships that we once had with materials, and they prevent us from attributing to them meanings that once endowed them with cultural and physical depth" [2]. Manzini's book was written as SMART materials and systems were only just beginning to emerge, so that his comments are largely directed at materials that were fixed once they left the factory or studio. SMART materials and systems offered materials and systems that had the capability to change, to sense and respond to changes in their environment.

SMART materials and systems began to emerge in architecture, military and sports in the early 1990s, bringing together a complex network of materials and technologies, each with their own signs, signifiers and systems. Lacking a singular provenance, they instead draw on many materials and technologies, each with its own distinctive provenance to the extent that for the first decade (and arguably longer), they were often dismissed as "solutions looking for problems" [3]. Although the sector now has a more clearly defined identity, the nature and the pace of the technologies and stakeholders means that it exists by necessity in a state of flux, of uncertainty.

It is the contention of this paper that uncertainty in this context is a positive. Used as a driver of innovation to propel the industry towards more timely commercialisation. Without innovation, there is little to interest the consumer, and without commercialisation, funding for new ideas cease to be justified. They are reliant on one another. In the first section of this paper the role of uncertainty in innovation is considered, drawing on literature and thought-leaders from innovation (Christensen), scenario planning (Wade) as well as the theories of Actor Network Theory (ANT) (Latour) and convergence(fusion) (Curren & Leker). The second portion goes on to apply these in the development of a toolkit that is designed to facilitate and harness the potential of uncertainty. The methodology builds on Wade's scenario building drawing on Curren and Leker's convergence(fusion) theory, demonstrating their combined potential in identifying future directions and achieving the timely commercialisation of SMART materials and systems.

2. The necessity of uncertainty

Uncertainty is not comfortable. The natural tendency of humans when faced with it is to try to eliminate it, to replace it with certainty as quickly as possible. Entrepreneur and business author Margaret Heffernan goes so far as to espouse uncertainty, pointing to humans' natural tendency to look for events and information that confirm our pre-existing world view: "We are very attracted to confirming data and we tend to marginalise and trivialise disconfirming data" [4]. The result the author argues is a natural gravitation towards a model of the world that will repel exactly the information that we need to be paying attention to.

In *The Innovator's Dilemma: When New Technologies Cause Great Firms to Fail* (2016), Clayton M. Christensen identifies two forms of innovation: sustaining and disruptive [5]. The former (that constitutes the majority) he classifies as offering improvements to product performance, so that it can be quite incremental. Disruptive innovation is less common, and introduces a very different value proposition to what was previously available. When new technologies emerge in a field it is very rare for the dominant company to retain its position over subsequent generations to the extent that "today's competitive advantage may become tomorrow's albatross unless strategists attune themselves to changes in underlying conditions"[ibid]. The need to consistently question and acknowledge uncertainties is challenging for industry, but for a disruptor there can be the added pressure that a sustaining innovation is not enough, it has to be disruptive once again.

Scenario planning at its most useful, is carefully constructed with an emphasis on anticipating rather than forecasting the future [6]. The reason being, that with so many variables, or unknowns, no single scenario could be guaranteed. Instead, scenario planning looks to identify an array of possible futures. Woody Wade in *Scenario planning – a field guide to the future* (2012) has identified ten steps

IOP Conf. Series: Materials Science and Engineering 1266 (2023) 012003 doi:10.1088/1757-899X/1266/1/012003

for scenario planning. In identifying driving forces, he uses a political, economic, societal and technological (P.E.S.T.) model that secures a clear starting point based on what has actually taken place and then moves forward to speculate on future possibilities and scenarios [ibid]. These steps culminate in the generation of four distinct scenarios or possible futures within a 2 x 2 matrix that has two axes and four quadrants. The ten steps are:

1. Framing the challenge
2. Gathering information
3. Identifying driving forces
4. Defining the future's critical "either/or" uncertainties
5. Generating the scenarios
6. Fleshing them out and creating story lines
7. Validating the scenarios and identifying further research needed
8. Assessing their implications and defining possible responses
9. Identifying signposts
10. Monitoring and updating the scenarios over time

Actor network theory (ANT) has a profound interest in how scientific knowledge is socially constructed. To consider this in depth both human and non-human actors need to be accounted for and by extension, activities and observation collected within the scientific laboratory and the wider world. Bruno Latour likens a network to a translation, and in this way offers more subtlety than a system [7]. A system implies a more rigid structure, while a network, or translation, allows for greater mobility. "Science has two faces: one that knows, the other that does not know yet" the need for a methodology that encompasses change and uncertainty is apparent [ibid]. It also points to the relevance of ANT to strategic foresight as a way of bringing together what is known, as well as the future that is as yet, unknowable.

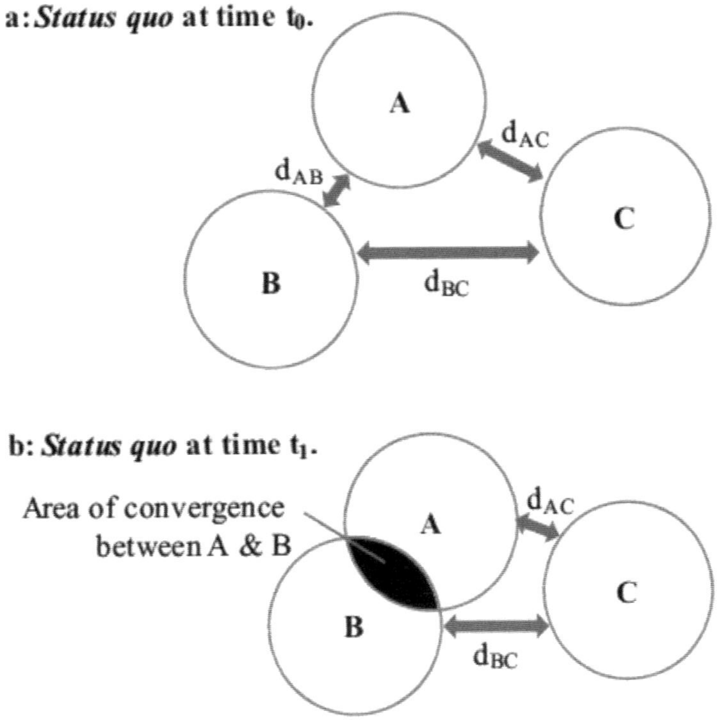

Fig.2. Curren and Leker's convergence(fusion) diagram

IOP Conf. Series: Materials Science and Engineering 1266 (2023) 012003 doi:10.1088/1757-899X/1266/1/012003

Clive-Steven Curren and Jens Leker's theory of convergence(fusion) refers to a blurring of boundaries between at least two hitherto distinct areas of science, technology, markets, or industries [8]. It allows for the creation of a new (sub-)segment in a new spot (the same spot) as a merger of (parts) of the old segments. The term is derived from the Latin "convergere," meaning inclining together as opposed to merging, indicating a retained separateness. This facilitates change, an increase of interchangeability and connectedness between the respective areas. In this way it is the uncertainty that is allowed for, even encouraged, that creates stability. Applying the principles of convergence(fusion) to SMART materials and systems it is possible to identify two distinct industries, the material (flexible) and the electronics (less flexible or hard) industry. This system and theory point to the importance of flux to create momentum and propel innovation and timely commercialisation.

3. The SMART Materials & Systems Toolkit

The challenges inherent in the innovation and commercialisation of SMART materials and systems are multiple, from the hybrid nature of the materials and technologies lacking a single distinctive provenance, to the very different actors required with each bringing their own histories and ways of working. The purpose of this toolkit is to offer a means of effective planning using what is already known for a future that is unknown.

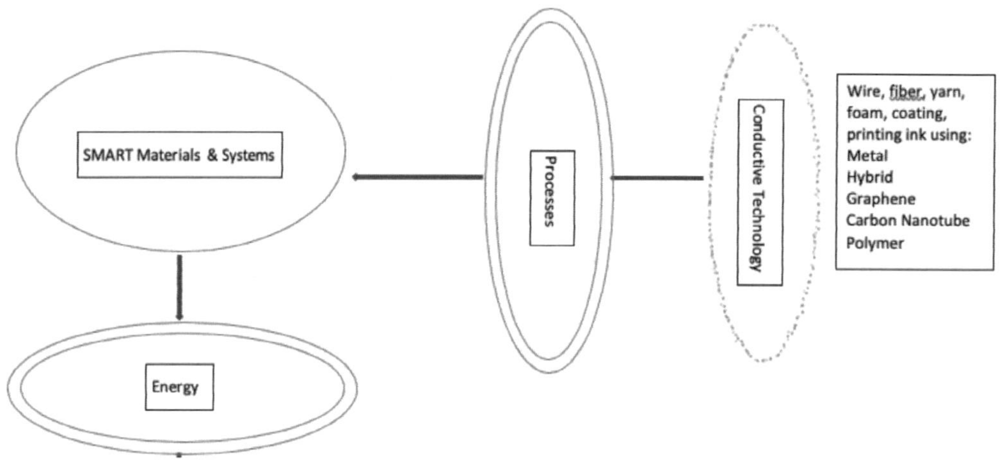

Fig.3. Object notation used to indicate a 'weighting' of their importance for each actor: mutable despite permanence (fluid, single line), a stable state (immutable mobile, double line) and unpredictability (fire, broken line)

The first portion of the toolkit is scenario building, typically using two timelines with the first date sufficiently near-term that some indicators can be reasonably identified. The second is long-term for a technology-related industry because of the rapid evolution of technologies. Employing ANT, the stakeholder is both human and non-human, and as internal and external actors, the stakeholder is not a fixed lens. Planning for the unknown must include the unanticipated future stakeholder who may also be a disruptor. The ten identifiable steps in scenario building proposed by Wade are taken as a starting point with the principles of convergence(fusion) introduced to allow for the hybrid nature of the materials, technology and systems.

The first step is framing the challenge, defining the stakeholders and the timelines to be applied. This is followed by information gathering, and the mapping of smart materials and systems within convergence(fusion) to provide a 'weighting' of their importance to each actor. The third stage is identifying driving forces gathering data and using a P.E.S.T. model. The fourth looks at critical extremes and the varying degrees of importance given to each in different circumstances. These are based on the driving forces identified in the P.E.S.T. analysis where each carries a potential impact, ranging from high to low. The level of impact and degree of uncertainty determines their critical uncertainty and from that identifies the two that are the most 'critically extreme.'

The fifth step is the development of four scenarios based on Wade's X-axis horizontal used to plot the uncertainties, with the intersecting Y-axis vertical to indicate the potential impact. In the sixth step, narratives are created with four storylines to generate possible alternative future plans. The seventh step is assessing implications and defining possible responses looking to allow for the greatest degree of flexibility. In the eighth step signposting is utilised to offer a process (flexible) rather than a fixed series of predictions. The ninth step is an outline for future monitoring and updating the scenarios using an annual P.E.S.T. analysis. The final tenth step is using the convergence(fusion) based scenario toolkit, proposing it as a starting point and itself subjected to revision and adaptation.

The value in this toolkit is predicated on the absence of a fixed outcome. The greatest challenge facing SMART materials and systems in industry and academia is not a disruptive technology or new actor, but instead the ability to embrace a network that is not fixed and where its stability and success is reliant on this.

4. Conclusion

The value of uncertainty has a long history in the sciences and problem-solving. Richard Feynman addressed it many times in his career, arguing for it as a necessity:

"All scientific knowledge is uncertain. This experience with doubt and uncertainty is important. I believe that it is of very great value, and one that extends beyond the sciences. I believe that to solve any problem that has never been solved before, you have to leave the door to the unknown ajar. You have to permit the possibility that you do not have it exactly right. Otherwise, if you have made up your mind already, you might not solve it."[9]

Certainty on the other hand points to something that is regarded as being a conclusive fact that is firm, and not open to further consideration. The door is closed. As the issue of commercialisation in SMART materials and systems is unresolved, this is problematic. Taking account of uncertainty, introducing it, managing and even embracing it offers a path to continuing innovation and bringing it to market in a timely manner.

References

[1] Braddock SE & O'Mahony M 1998 *TechnoTextiles* (London: Thames and Hudson) p 6
[2] Manzini E 1989 *The Material of Invention – materials & design* Translator Anthony Shugaar (Cambridge Massachusetts: Institute of Technology MIT Press) p 31
[3] van Hinte E 1998 *Smart Design: report on the 2nd smart materials and systems workshop* (Amsterdam: Netherlands Design Institute NDI) p 11
[4] Heffernan M 2020 *Interviewed by Helen Lewis* The Spark BBC Radio 4 1 April 2020 Available at: https//bbc.co.uk/sounds/play/m000gsm8 Accessed 26 August 2022
[5] Christensen CM 2016 *The Innovator's Dilemma: When New Technologies Cause Great Firms to Fail* 3rd Edition. Boston: Harvard Business Review Press
[6] Wade W 2012 *Scenario planning – a field guide to the future* Holboken New Jersey: Wiley p 20, 36
[7] Latour B 1987 *Science in action: how to follow scientists and engineers through society* Cambridge, Massachusetts: Harvard University Press p 3
[8] Curren CS & Leker J 2011 *Patent indicators for monitoring convergence - examples from NFF and ICT* (Technological Forecasting & Social Change 78 2011) pp 256-273
[9] Feynman R 2005 *The meaning of it all: thoughts of a citizen-scientist* New York: Basic Books p 13 Available at: http://www.inf.fu-berlin.de/lehre/pmo/eng/Feynman-Uncertainty.pdf Accessed 28 August 2022

Flexible piezoelectric nanocomposites based on PVDF-HFP/PLA blend doped PZT

K Oumghar [1,2*], N Chakhchaoui [1,3,4], A Eddiai [2], O Cherkaoui [1] and M Mazroui[2]
[1]REMTEX Laboratory, Higher School of Textile and Clothing Industries (ESITH), Casablanca, Morocco
[2]Laboratory of Physics of Condensed Matter (LPMC), Faculty of Sciences Ben M'Sik, Hassan II University, Casablanca, Morocco
[3]Laboratory Nanotechnologies and Nano Systems, LN2, CNRS, University of Sherbrooke, 3IT (Interdisciplinary Institute of Technological Innovation), Sherbrooke, QC, Canada
[4]Institut Interdisciplinaire d'Innovation Technologique (3IT), Université de Sherbrooke, 3000 Boulevard Université, Sherbrooke, J1K OA5 Québec, Canada
* khadija.oumghar-etu@etu.univh2c.ma

Abstract. In the present work, we studied the effect of adding PZT nanoparticles to PVDF-HFP/PLA matrix on piezoelectric properties of blend-based nanocomposites. Polyvinylidene fluoride hexafluoro-propylene /Polylactic acid /Lead titanate zirconate (PVdF-HFP/PLA/PZT) films nanocomposites were prepared by solvent casting technique using different percentages of PZT Nano fillers. The different samples were characterized by Polarized optical microscope (POM), Fourier transformed infrared (FTIR) and Thermogravimetric analysis (TGA).
POM images indicate uniform distribution of PZT Nano fillers in the PVdF-HFP/PLA matrix. FTIR analysis reveals the appearance of β-phase in nanocomposites and the enhancement of their piezoelectric properties. These electroactive nanocomposites thin films are a potential candidate for the piezoelectric Nano generators, energy storage devices and energy harvesting applications.

1. Introduction

The capacity of some materials, notably ceramics and crystals, to create an electrical voltage from applied mechanical stress and vice versa is referred as piezoelectricity. Piezoelectric inorganic materials such as quartz and ceramic have been employed in electromechanical transduction since their discovery. Because of the difficulty of producing these piezoelectric inorganic materials at high temperatures, toxicity, and a lack of flexibility, polymeric materials have become the material of choice [1].
In this work, we are prepared a binary polymer blend by addition of poly lactic acid (PLA) as a polar polymer and PZT as piezo ceramic Nano fillers to poly PVdF-HFP matrix. This films were prepared by solvent casting method. Both PVdF-HFP and PLA have good piezoelectric properties [2].
These nanocomposites have capacity to transform mechanical energy into electrical energy to be the great candidates for smart textiles applications [3].
The effects of blending nanoparticles on morphological properties, crystallization and melting behavior, were characterized by polarized optical microscope (POM), Fourier transform infrared (FTIR) and Thermogravimetric analysis (TGA). The POM results showed a uniform distribution of PZT nanoparticles in the polymer matrix.
Spectroscopic studies by Fourier transform infrared spectroscopy (FTIR) and thermal studies (TGA) analysis confirmed the improvement in electroactive polar phases of PVdF-HFP, thus revealing the successful intermolecular interaction between CH2 and CF2 group of PVdF-HFP and -C=O group of polylactic acid (PLA) leading to good compatibility between both the polymers, and showed that β-phase increased as a result of blending while α-phase was decreased. The addition of PZT in mixtures showed a synergistic effect on polymorphic β-phase and developed its potential applications.

2. Materials and Methods

The materials used are: Lead Titanate-Zirconate (PZT) nanoparticles, Polylactic acid (PLA), Poly (vinylidene-fluoride hexafluoropropylene) (PVdF-HFP) and Tetrahydrofuran (THF) as a solvent.

The films nanocomposites were prepared by the solvent-casting method.
First the Polyvinylidene fluoride-hexafluoropropylene (PVdF-HFP) and Polylactic acid (PLA) were separately dissolved in Tetrahydrofuran (THF). After the two solutions are dissolved, they are mixed to prepare a PVdF-HFP/PLA solution. On the other hand, different amounts of lead titanate zirconate (PZT) nanoparticles are prepared by dispersing the nanoparticles of PZT (powder) in Tetrahydrofuran (THF) using ultrasonication. Then, the two solutions are mixed. Afterwards, the mixture is poured and dries to form the PVdF-HFP/PLA/PZT nanocomposites films.

3. Results and discussion

3.1. Polarized optical microscope

The morphological characterization of nanocomposite films was studied by polarized optical microscope.
Figure 1 illustrates the polarizing optical microscope (POM) images of PVDF-HFP/PLA/PZT blends with different percentages of PZT. As shown in Figure 1, the PZT Nano fillers distributes homogeneously inPVDF-HFP matrix with addition of PLA. The homogeneous dispersion of PZT in PVDF-HFP/PLA have significant influence on the piezoelectric properties of nanocomposites [4]. The result indicates that the interface strength between PZT and PVDF-HFP matrix are enhanced effectively by introducing PLA component. Accordingly, these reinforced interface interaction improves the dispersion of PZT in PVDF-HFP matrix.

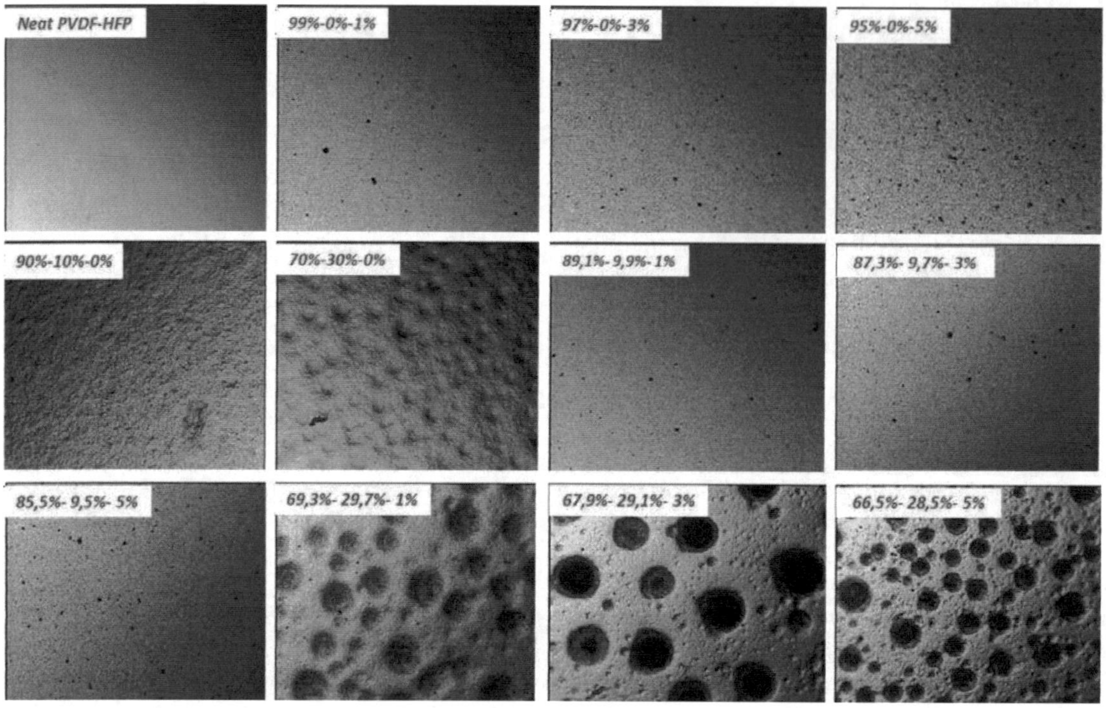

Figure 1. Polarized optical microscope of PVDF-HFP/PLA/PZT nanocomposites.

3.2. Fourier transform infrared spectroscopy

Figure 2 presents the FTIR spectra of the pure PVdF-HFP, pure PLA, PVdF-HFP/PLA blends and PVdF-HFP/PLA/PZT nanocomposites. The vibrational peaks at 1176 cm-1 and 1753 cm-1 are respectively corresponding to the –C-H and –C=O groups of PLA. The characteristics vibrational peaks observed at 1071, 1176 and 1400 cm^{-1} are corresponding to bending vibration C– C, swinging vibration CH_2 and motion vibration CF_2 groups of PVdF-HFP, respectively [5].
The characteristic peaks disclosed at 764 cm^{-1}, 796 cm^{-1} and 1071 cm^{-1} correspond to the crystalline α-phase of PVdF-HFP, And the vibrational peaks of piezoelectric crystalline β-phase are observed at 1431 cm^{-1} [6]. Figure 2 shown that with the addition of PLA to PVdF-HFP, the peak at 1431 cm^{-1} of β-phase has not changed. After adding PZT, we are observed that β-phase increased, while α-phase at 1071 cm−1 was decreased indicating that PLA and PZT Nano fillers promoted the formation of β-phase in nanocomposites.

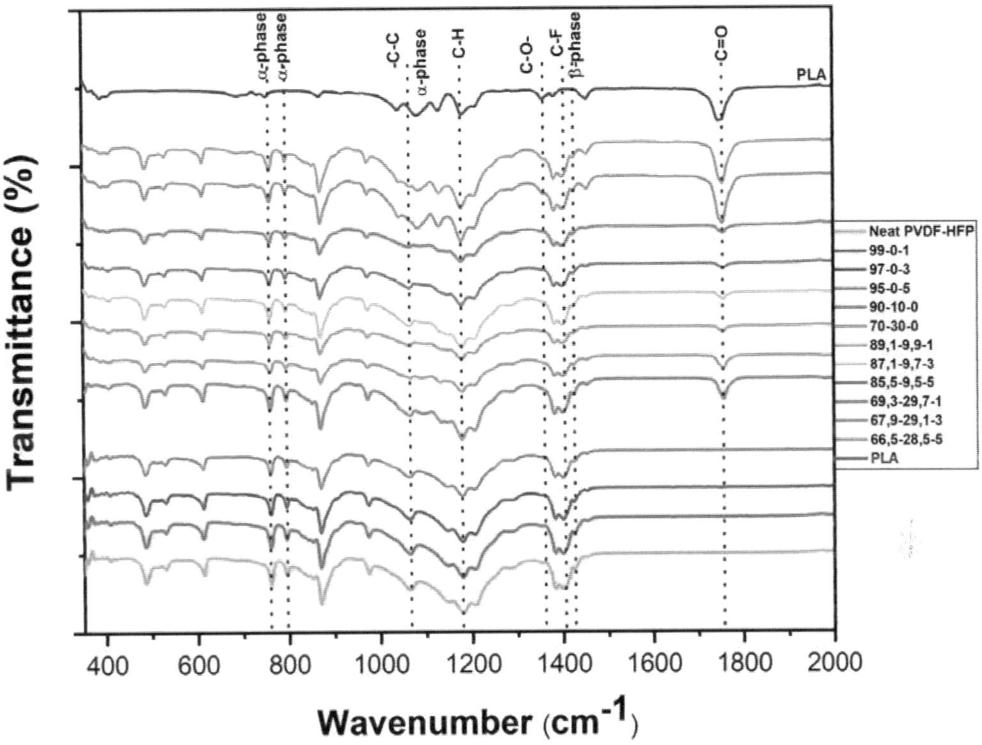

Figure 2. Infrared spectroscopy of PVDF-HFP/PLA/PZT nanocomposites.

3.3. Thermogravimetric analysis

Because PZT ceramics have good thermal stability at the processing temperatures utilized for most elastomers and engineering plastics, the thermal stability of PZT-based polymer nanocomposites is of interest [7,8]. In nitrogen conditions, the thermal degradation of extruded PZT-based PVDF-HFP/PLA nanocomposites blend was evaluated, with TGA curves shown in Figure 3. The results reveal that increasing the PZT concentration in PVDF-HFP/PLA nanocomposites results in a considerable increase in thermal stability, particularly at the commencement of degradatio. Figure 3 depicts a single-step degradation process for pure PVDF-HFP, pure PLA, and PVDF-HFP/PLA/PZT nanocomposites at nitrogen environment.

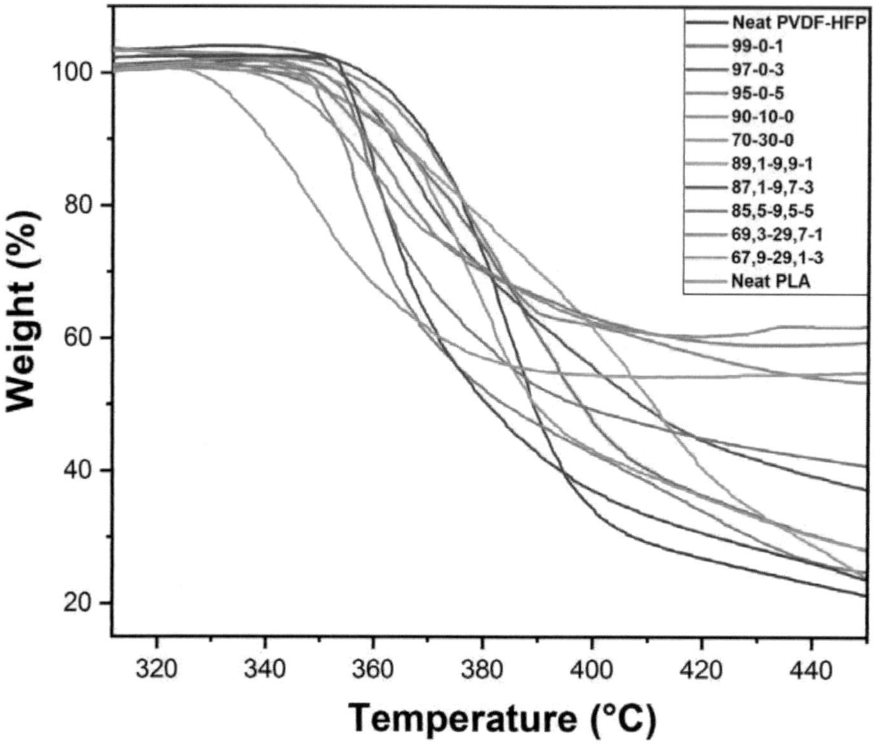

Figure 3. Thermogravimetric analysis of PVDF-HFP/PLA/PZT nanocomposites.

4. Conclusion

PVDF-HFP/PLA/PZT blends were developed to improve the piezoelectric properties of nanocomposites by enhancing the electroactive β-phase.
The results of morphological, structural and thermal characterizations analyzed by polarized optical microscope (POM), Fourier transform infrared spectroscopy (FTIR) and Thermogravimetric analysis (TGA) indicate the strong interaction between the PZT Nano fillers and the PVDF-HFP/PLA mixture matrix. The nanocomposite exhibits a high piezoelectric effect by the appearance of β-phase and the disappearance of α-phase due to the good dispersion of thePZT Nano fillers.

References:

[1] Eddiai A, Meddad M, Farhan R, Mazroui M, Rguiti M, et Guyomar D, 2019, *Superlatticesand Microstructures,* 127, pp 20-26.

[2] Chakhchaoui N, Jaouani H, Ennamiri H, Eddiai A, Hajjaji A, Meddad M and Boughaleb Y, 2019 *Journal of Composite Materials* vol 53 (24) pp 3349-3361

[3] Thakur P, Kool A, Bagchi B, Das S, and Nandy P 2015 *Physical Chemistry Chemical Physics* vol 17, (2) pp 1368-1378.

[4] Cai X, Lei T, Sun D, and Lin L 2017 *RSC advances* vol 7 (25) pp 15382-15389.

[5] Busolo T, Szewczyk P K, Nair M, Stachewicz U and Kar-Narayan S 2021 *ACS applied materials & interfaces* vol 202113 (14) pp 16876-16886.

[6] Oumghar K, Chakhchaoui N, Farhane R, Eddiai A, Meddad M, Cherkaoui O and Van Langenhove L 2020 *IOP Conference Series: Materials Science and Engineering* Vol 827 No. 1 p 012034.

[7] Oumghar K, Chakhchaoui N, Farhane R, Eddiai A, Meddad M, Cherkaoui O and Van Langenhove L 2020 *IOP Conference Series: Materials Science and Engineering* Vol 948 No 1 p 012024.

[8] Bouiri E. M, Farhan R, Chakhchaoui N, Oumghar K, Denktas C, Eddiai A, and Omari L. E. H. 2022 *The European Physical Journal Applied Physics* Vol 97 64.

Investigation of conductive printed thermoplastic polyurethane nanofibers to detect the clogging of air filters

Parian Mohamadi*[1], Elham Mohsenzadeh[1, 2] Cedric Cochrane[1], Vladan Koncar[1]

[1] Univ. Lille, ENSAIT, ULR 2461 - GEMTEX - Génie et Matériaux Textiles, F-59000 Lille, France
[2] Junia, F-59000 Lille, France

E-mail: parian-sadat.mohamm@ensait.fr

Abstract. Recently, air pollution attracted many worries because of its high number of deaths per year. To solve the problem, the industries are trying to fabricate the giant air filtration system for public areas. However, the clogging of air filters should be detected in real-time to change or clean them. E-textile is a very fascinating field, which is often used in medical, safety, military and clogging detection applications. These components are integrated into soft textile materials according to their usage requirements. One of the most attractive textile structures is the nanofibers due to their advantageous properties such as porosity, lightweight, and high surface area. To have conductive nanofiber-based membrane sensors, two in situ electrical conductivity principles using conductive particles and surface conductivity, such as immersion and printing methods are recommended. In this research, the thermoplastic polyurethane (TPU) nanofibers' membranes are produced using an electrospinning system and the carbon ink was printed on the surface of nanofibers to apply in textile sensors applications. SEM images showed the uniform structure of the nanofibers and the porosity of the system even after printing. The electromechanical properties of printed membranes demonstrated the change of electrical resistance under stretch. Conclusively, these conductive membranes could be employed as strain sensors to detect the small changes in the output airflow indicated the possible clogging of air filters.

1. Introduction

Nowadays, air filters have an important role in public areas such as swimming pools and shopping malls for improving the air quality by transferring out the particles of different sizes produced by various activities. High efficiency and low-pressure droplets can be mentioned as the quality of a proper filter, for removing the dust and better air passing through the filter, respectively. These properties should be preserved during the use of the system [1]. Clogging is one of the most significant matters of the fibrous filters caused by the accumulation of small particles in the pores, which industries are trying to solve. However, the first step is the detection of the clogging which indicates that a filter has to be replaced or cleaned [2]. Over the years, textile sensors including pressure, optical, strain, humidity, temperature, and chemical sensors are strongly used in various fields of applications such as medicine, military, aerospace, and electronics sectors. They could also be used to notify the clogging of a filter, considering different conditions including humidity, temperature, etc., before and after the filter. Figure 1 indicates the textile sensors in a filtration system.

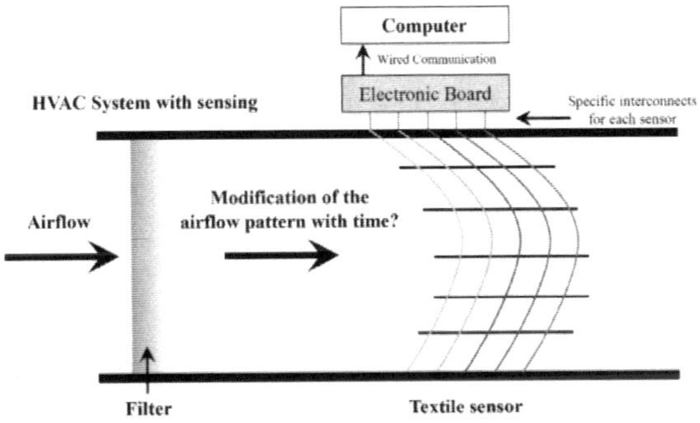

Figure 1. Textile sensors in a filtration system.

The structure of these sensors can be based on nanofibers; nonwovens, films, wrapped fibres, and yarns, and they could also be produced by coating, immersion, electrospray, and electrospinning techniques [3]. Zhu et al. fabricated the highly stretchable TPU/CNT composite with a 3D porous skeleton structure which could be used as strain, and pressure sensors as they can sense the tactile stimuli of stretching and compression. This system showed an ultralow strain detection limit (0.01%) and ultrawide strain sensing range (0.01–900%) [4]. The nanofibers have interesting structures due to their advantages, such as a high surface area and low weight [5]. They also have very interesting mechanical properties and can detect a very small variation in the airflow before and after the filter. Therefore, they allow the sensing of the clogged areas. Electrospinning is a cost-effective and simple method of fabricating nanofibers compared to template, drawing, phase separation, and self-assembly methods [6]. The electrospinning mechanism is shown in Figure. 2. Electrospun mats are produced by applying a high-voltage electrostatic field on the jet of natural or synthetic polymer solutions [7]. Also, using synthetic polymers could improve the mechanical properties of the sensing system and enhance its sensitivity.

TPU is one of the synthetic polymers, which could be easily electrospun into the fibrous membrane and it shows stretchability and flexibility properties [8]. Wang et al., produced TPU nanofibers using two different collector speeds of 100 and 200 rpm. To have a conductive layer, they immersed the samples into carbon black (CB) solutions under ultrasonication. All samples showed a change in electrical resistance against stretch that could be because of the tunnelling mechanism between the particles [9].

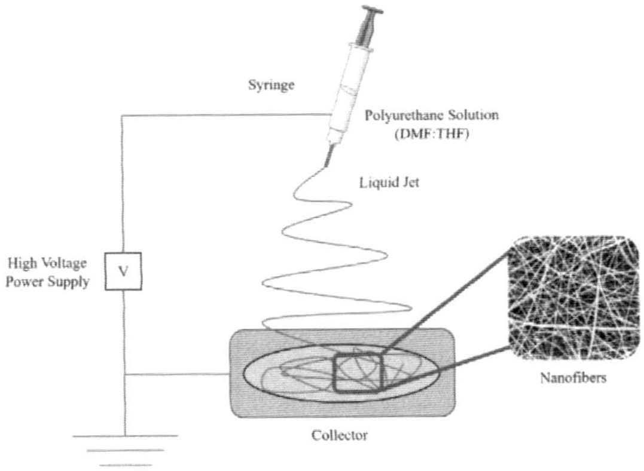

Figure 2. The electrospinning mechanism.

The goal of this study is the fabrication of polyurethane nanofibers using DMF/THF as solvents followed by printing the carbon paste ink on the surface of membranes to have conductive parts. To achieve high elasticity and sensitivity, the parameters of electrospinning and printing were optimized, respectively. Finally, as the membranes made of nanofibers are electrically conductive, the variations in the airflow will be detected by a simple measurement of the electrical resistance within the Wheatstone bridge. Additionally, these properties permit the detection of clogging in air filters of the filtration system.

2. Materials and methods

Electrospinning setup FlUIDNATEK/LE-50 model was from Bioinicia Company, Spain. Thermoplastic polyurethane (Laripur 5760 with MW:100,000-200,000 Da) was bought from Coim Group, Italy. N,N-Dimethylformamide (DMF: 98% purity) and Tetrahydrofuran (THF 98% purity) were provided from Sigma-Aldrich. DM-CAP-2100S (carbon paste) with a conductivity of < 20 Ω/sq was purchased from Dycotec materials, UK. Finally, the Voltera V-One PCB Printer machine (Voltera company, Canada) was used to print the ink on the membranes.

To produce TPU electrospun membranes, the TPU pellets were dissolved in DMF:THF (1:1 v/v) and mixed for 5 hours using a magnetic stirrer to obtain 11 wt.% of the polymeric solution. Then, the polymeric solution was transferred into a syringe and the electrospinning parameters were performed at 12 kV voltage, 1.5 mL/hr feed rate, 15 cm nozzle to the collector distance, and 200 rpm speed of cylindric collector. Then, the carbon black ink was printed on the surface of nanofibers using a feed rate of 5 μm in addition to 0.01 and 0.1 mm of height (needlepoint to nanofibers surface distance) as printing parameters. The printing pattern was designed by applying Kicad software as demonstrated in Figure. 3. Finally, copper wires were connected to samples using carbon ink for electrical measurements and the printed nanofibers were kept in an oven at 80 °C for 20 min to dry the ink.

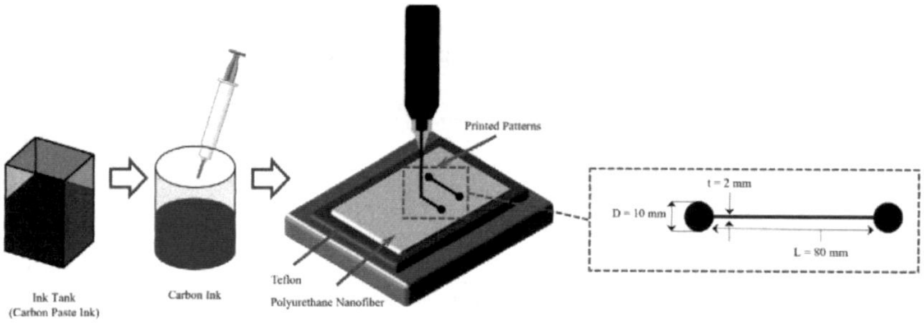

Figure 3. The printing method on the membranes.

3. Characterization

3.1 Scanning Electron Microscope (SEM)

To verify the morphology of nanofibers and the porosity, SEM images were taken using Phenom ProX, Thermo Fischer Scientific, US device. The average diameter of prepared nanofibers was determined using Digimizer software (mean of 50 measurements).

3.2 Electromechanical properties

Rectangular-shaped samples with a length of 80 mm and a width of 10 mm were prepared according to the printed patterns to investigate the mechanical properties of the samples with the MTS criterion machine (Model 43). The crosshead speed was set at 10 mm/min, and at least three specimens were tested for each sample as well as the determination of the maximum force and elongation at break. To evaluate the electromechanical properties, copper wires were connected to both sides of the printed part, with carbon ink to guarantee good contact between the electrodes (the copper wires) and electrospun nanofibers composite strip. Thereafter, the initial electrical resistance of the specimens was measured using a resistance box and the real-time voltage changes were recorded with the gradual stretching of the specimens. Finally, the sensing responsivity, GF, which is defined as the ratio of $\Delta R/R$ to $\Delta L/L$, was reported.

4. Results and Discussions

4.1. Morphology of nanofibers
As it is demonstrated in Figure. 4, the SEM images confirm that the electrospinning parameters were well optimized and the beads-free, uniform, and homogenous nanofibers were produced. The mean diameter of nanofibers was measured 796 ±79 nm.

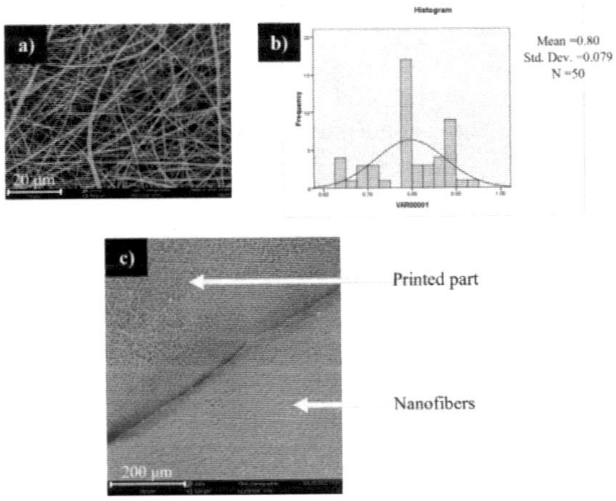

Figure 4. The SEM images of a) TPU membrane b) distribution of diameters, and c) printed TPU membrane.

The images were taken after printing to see the blockage of the pores and according to the results, the blockage was observed just in the printed region. As a result, the porosity of nanofibers was preserved, therefore, they could be used in filtration systems by allowing the air to pass through the filter.

4.2 Mechanical properties
Figure 5 and Table 1 represent the average force-strain curve and maximum force and elongation at the break of neat TPU and printed TPU membranes, respectively. In many studies, TPU is recommended as a flexible and elastic polymer for strain sensor applications. He et al. produced TPU/PCL nanofibers treated by PEDOT:PSS/CNT that showed more than 400% of strain and could display the capability of detecting human motions [10]. Shehata et al. reported that adding 30% of TPU to PVDF nanofibers improved the elasticity of membranes up to 97% for wearable electronics [11].

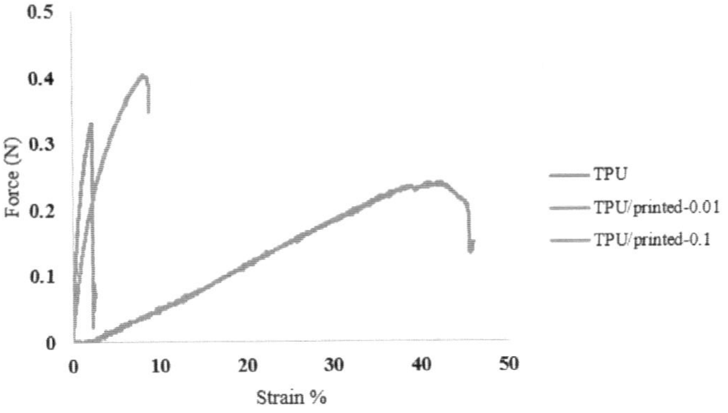

Figure 5. Force-strain curves of TPU membrane, TPU/printed-0.01, and TPU/ printed-0.1.

Table 1. Mechanical properties of TPU membrane, TPU/printed-0.01, and TPU/ printed-0.1.

Samples	Max force at break (N)	Max elongation at break (%)
TPU	0.23 ± 0.02	43.03 ± 2.04
TPU/printed-0.01	0.32 ± 0.03	2.2 ± 1.90
TPU/printed-0.1	0.39 ± 0.04	8.68 ± 0.49

The presence of a printing layer decreased the strain of TPU from 43% to 2% and 8% for TPU/printed-0.01 and TPU/printed-0.1, respectively. However, an increase in force was observed after printing but the difference between samples was not significant. The reason could be the rigidity of carbon ink that restricted the elongation of the membrane, and the break was detected in the printing region for all samples. It explains that the printing parameter such as feed rate and height determine a key role in the final properties. As shown in Figure. 5, by increasing the distance between the needle of printing and membrane from 0.01 mm to 0.1 mm, the strain increased from 2% to 8%. As a result, the printing parameters should be optimized to obtain the optimal mechanical properties of these strain sensors.

4.3 Electromechanical properties

The sensing performance of the sensor when stretched was displayed in Figure. 6. The gauge factor (GF) is defined as $(\Delta R/R_0)/\varepsilon$ (where R_0, ΔR, and ε show the initial resistance, resistance change, and elongation respectively). Reducing the initial resistance (R_0) and increasing the resistance change is the important factor for the sensitivity of strain sensors.

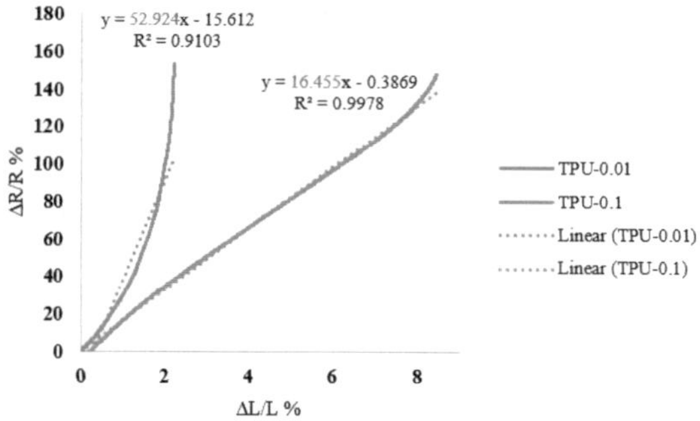

Figure 6. Tensile sensing curve of TPU membrane, TPU/printed-0.01, and TPU/ printed-0.1.

According to the curves shown in figure 6, by applying the strain, the resistance changes increased, and the graphs indicated linear behaviour for two printed samples. The GF (shown red in the curves) was calculated at 52.92 and 16.45 for TPU/printed-0.01 and TPU/printed-0.1, respectively. Although by reducing the distance between the printing needle and the membrane, the elongation decreased, higher sensitivity was observed because of the presence of a thicker carbon layer on the surface of the membranes. Compared to relative research, the GF of strain sensors was less than 30 for up to 40% strain. Therefore, the strain sensor prepared in this study has superior sensitivity to many similar sensors [12-14]. As a result, this system could act as a strain sensor and demonstrate a piezoresistive effect by changing the electrical resistance under stretching. In addition, contrary to other works, here the printed TPU membranes showed higher sensitivity even under 2% strain [15].

5. Conclusions

In this research, the flexible TPU membranes were fabricated using the electrospinning method. Then, the carbon ink was injected onto the surface of electrospun membranes to print a straight-line pattern with a length and width of 8 mm and 1 mm, respectively. The electrospinning and printing parameters were optimized and the SEM images showed the beads-free, uniform, and homogenous structure of nanofibers and confirmed the presence of the porosity even after printing. TPU membrane without conductive parts indicated up to 43% strain while this amount decreased to 2% and 8% according to the needle distance of the printing machine. However, the maximum force at break improved by adding a printing layer due to the rigidity of carbon ink. The tensile sensing test represented the resistance change of samples under stretch and the GF was calculated at 52.92 and 16.45 for TPU/printed-0.01 and TPU/printed-0.1, respectively. Thus, the thicker layer of carbon ink followed by reducing the needle distance of printing equipment could enhance the sensitivity of samples and the resistance change even under 2% strain.

In conclusion, this system could be useful in strain sensor applications for air filters' clogging detection because of its exciting properties such as elasticity, porosity, and sensitivity against stretch. In the future, the printing parameters such as feed rate and height of needle should be optimized to achieve proper mechanical and electromechanical properties and distinct printing patterns could be worked to compare and obtain the sensitivity in different directions.

References

[1] G. Liu, M. Xiao, X. Zhang, C. Gal, X. Chen, L. Liu, S. Pan, J. Wu, L. Tang, D. Clements-Croome, A review of air filtration technologies for sustainable and healthy building ventilation, Sustainable cities and society 32 (2017) 375-396.

[2] C.-B. Song, H.-S. Park, K. Lee, Experimental study of filter clogging with monodisperse PSL particles, Powder technology 163(3) (2006) 152-159.

[3] L.M. Castano, A.B. Flatau, Smart fabric sensors and e-textile technologies: a review, Smart Materials and structures 23(5) (2014) 053001.

[4] G. Zhu, H. Li, M. Peng, G. Zhao, J. Chen, Y. Zhu, Highly-stretchable porous thermoplastic polyurethane/carbon nanotubes composites as a multimodal sensor, Carbon (2022).

[5] M. Shokrollahi, S.H. Bahrami, M.H. Nazarpak, A. Solouk, Multilayer nanofibrous patch comprising chamomile loaded carboxyethyl chitosan/poly (vinyl alcohol) and polycaprolactone as a potential wound dressing, International Journal of Biological Macromolecules (2020).

[6] F. Haghdoost, S.H. Bahrami, J. Barzin, A. Ghaee, Preparation and characterization of electrospun polyethersulfone/polyvinylpyrrolidone-zeolite core–shell composite nanofibers for creatinine adsorption, Separation and Purification Technology 257 (2021) 117881.

[7] S.C. Coelho, B.N. Estevinho, F. Rocha, Encapsulation in food industry with emerging electrohydrodynamic techniques: Electrospinning and Electrospraying-A review, Food Chemistry (2020) 127850.

[8] Y. Wang, W. Li, Y. Zhou, L. Jiang, J. Ma, S. Chen, S. Jerrams, F. Zhou, Fabrication of high-performance wearable strain sensors by using CNTs-coated electrospun polyurethane nanofibers, Journal of Materials Science 55 (2020) 12592-12606.

[9] X. Wang, X. Liu, D.W. Schubert, Highly sensitive ultrathin flexible thermoplastic polyurethane/carbon black fibrous film strain sensor with adjustable scaffold networks, Nano-Micro Letters 13(1) (2021) 1-19.

[10] X. He, J. Shi, Y. Hao, L. Wang, X. Qin, J. Yu, PEDOT: PSS/CNT composites based ultra-stretchable thermoelectrics and their application as strain sensors, Composites Communications 27 (2021) 100822.

[11] N. Shehata, R. Nair, R. Boualayan, I. Kandas, A. Masrani, E. Elnabawy, N. Omran, M. Gamal, A.H. Hassanin, Stretchable nanofibers of polyvinylidenefluoride (PVDF)/thermoplastic polyurethane (TPU) nanocomposite to support piezoelectric response via mechanical elasticity, Scientific Reports 12(1) (2022) 1-11.

[12] L. Liu, S. Chen, A. Xu, G. Cai, Manufacturing high sensitive strain sensor of polyurethane nanofiber Mat/AgNWs by simple dip-dry method, Fibers and Polymers 21(2) (2020) 359-365.

[13] Y. Jiang, Y. Chen, W. Wang, D. Yu, A wearable strain sensor based on polyurethane nanofiber membrane with silver nanowires/polyaniline electrically conductive dual-network, Colloids and Surfaces A: Physicochemical and Engineering Aspects 629 (2021) 127477.

[14] L. Tong, X.-X. Wang, X.-X. He, G.-D. Nie, J. Zhang, B. Zhang, W.-Z. Guo, Y.-Z. Long, Electrically conductive TPU nanofibrous composite with high stretchability for flexible strain sensor, Nanoscale Research Letters 13(1) (2018) 1-8.

[15] L. Lin, L. Wang, B. Li, J. Luo, X. Huang, Q. Gao, H. Xue, J. Gao, Dual conductive network enabled superhydrophobic and high performance strain sensors with outstanding electro-thermal performance and extremely high gauge factors, Chemical Engineering Journal 385 (2020) 123391.

Inkjet printing of silver conductive ink on textiles for electronic applications: impact of ink formulation on electrical performances of the ink

Abdelkrim Boumegnane[1,2,4], Assia Batine[1,2], Ayoub Nadi[2,4], Abdelouahed Dahrouch[3], Abdelhamid Stambouli [3], Omar Cherkaoui[2], Mohamed Tahiri [1]

[1]Organic Synthesis, Extraction and Valorisation Laboratory (SOEV) faculty of Sciences Ain Chock, Hassan II University. Km 8 Route d'El Jadida, B.P 5366 Maarif 20100. Casablanca, Morocco.

[2]Laboratory of Research on Textile Materials (REMTEX), Higher School of Textile and Clothing Industries (ESITH), 20230 Casablanca, Morocco.

[3] Royal Gendarmerie Forensic Science Institute, Morocco

[4]Author to whom any correspondence should be addressed:
boumegnanechimie.96@gmail.com; nadi@esith.ac.ma

Abstract. Printed electronics technology is one of the most dynamic in the world, allowing for the low-cost fabrication of electronic networks on textile substrates using the inkjet printing technique which is commonly used in various industries. In the field of formulation of conductive inks, silver nanoparticles are generally used as precursors that confer electrical conductivity to the printed patterns. In the present work, we synthesized silver nanoparticles by an ecological reduction method and then dispersed them in a PEG/Glycerol mixture to prepare a conductive ink. The silver nanoparticles were characterized by X-ray diffraction (XRD), as well as the morphology of the printed silver tracks was characterized by SEM. The developed ink was then successfully printed on a piece of pre-treated cotton fabric to produce flexible electronic components on the textile.

1. Introduction

Inkjet printing is considered a feasible manufacturing process for low-cost, advanced integrated electronic devices. Various electronic components can be developed with this technology, such as solar cells [1], RFID tags [2], transistors [3], flexible displays [4], extensible heaters [5], and sensors [6]. Due to its low cost, reliability, ease, and speed of operation, printed electronics have attracted growing attention to material and system aspects [7]. Inkjet printing is an additive manufacturing technology for depositing a conductive layer on various substrates, such as textiles, with high resolution and design freedom for textile-based electronic applications, giving rise to the idea of E-Textile [8]. Printable conductive ink has been the subject of many research studies as an important component of inkjet printing technology. Due to their strong conductivity, silver nanoparticles are the most commonly used for conductive ink formulation. It is important to note that silver and oxidized silver have the best electrical conductivity, making them widely used to make conductive patterns for electronic devices [9]. Here we report the printing of silver ink on textiles for wearable electronic applications. We first synthesized the silver nanoparticles, then, we optimized the suspension matrix by formulating a range of inks with different amounts of glycerol. Finally, the optimized ink was printed on surface-pretreated textiles for highly conductive wearable E-textile applications.

2. Experimental section

2.1. Material

Cotton fabric (89.2g/cm^2) was designed for printing. All the reagents were purchased and used without additional purification. Strach was purchased from a domestic seller. Silver nitrate ($AgNO_3$; ≥99%) was purchased from VWR chemicals, sodium carbonate (Na_2CO_3; ≥98%) was purchased from Fluka Chemika, Sodium Hydroxide (NaOH), PEG (1000g/mol), and glycerol were purchased from Sigma Aldrich.

2.2. Preparation of silver conductive ink

Silver nanoparticles were prepared by simple green reduction processes according to the procedure described in our previous work [10]. For the preparation of the nanoparticle suspension solution, glycerol was mixed with PEG and distilled water. The obtained nanoparticles were dispersed in the suspension solution to prepare the conductive nanoparticle ink. The silver nanoparticles were dispersed by sonication to obtain a 2.5 wt% silver nanoparticle ink and used for printed flexible circuits. This percentage was targeted according to the results of the procedure described in our previous work, in which we found that the optimal percentage of AgNPs is 2.5%wt [10].

2.3. Cotton fabric pre-treatment

The treatment of cotton fabric with thermoplastic polyurethane was carried out using a coating technique to smooth the textile surfaces. The fabric sample was treated with a viscous polyurethane solution with a dilution of 50%. The curing of the treated fabrics was carried out at 150 °C for 3 min.

2.4. Printing of AgNPs ink

In this step using a syringe, different conductive lines of silver nanoparticle ink with a concentration of 2.5 wt% are applied to the PU-treated cotton, followed by a drying step at 100 °C for 3 minutes.

2.5. Characterization.

Crystal structure analysis of silver nanoparticles was performed by X-ray diffraction (Bruker AXS D4) recorded at a 2θ scale from 10° to 80°. Rheological characterization was performed using a HAAKE RHM01-RD rheometer, all measurements were repeated three times at 25 °C. The surface morphology of the printed cotton was studied by scanning electron microscopy. The observations were made with Quanta 650 Thermo Fisher Scanning Electron Microscopy equipped with an EDS detector (EDAX Instruments Apollo SDD). An MTX 3292 two-rod multimeter, was used to measure the electrical resistance R (Ω) of the textile samples printed with conductive ink.

3. Results and discussions.

3.1. XDR analysis.

To investigate the influence of the suspension solution on the rheological and conductive properties, different conductive inks were prepared by adding silver nanoparticles as conductive fillers to the suspension solution. **Figure 1** shows the XRD pattern of silver nanoparticles, five characteristic peaks are observed at 38.08°, 44.26°, 64.42°, 77.36°, and 83.56°, corresponding to the diffraction pattern of the (111), (200), (220), (311) and (331) crystal plane of face-centered cubic (FCC) structure of silver, respectively [11].

Using the Scherrer equation (1), the crystallite size of the nanoparticles was determined.

$$D = \frac{k \times \lambda}{\beta \times cos\theta} \quad (1)$$

Where D is the crystallite size, k is the Scherrer constant (0.9), λ is the beam wavelength, θ is the diffraction angle, and β is the total width at half height of the peak. Using this formula, the dominant particle size is 32.29 nm (**Table 1**).

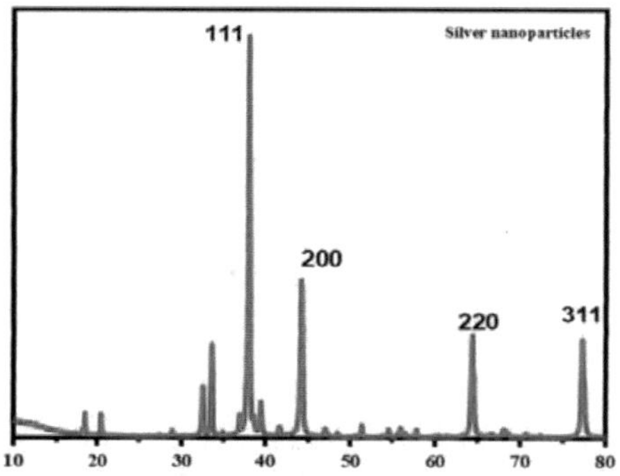

Figure 1: X-ray diffractogram of AgNPs.

Table 1: The full range of particle size measurements.

Peak position 2θ (°)	hkl	β (°)	Dp (nm)	Dp Average (nm)
38.08	111	0.25579	34.33	32.29
44.26	200	0.30564	29.32	
64.42	220	0.35117	27.93	
77.36	311	0.4056	26.22	

3.2. The viscosity meseaurment

The viscosity of the suspension matrix is a critical property parameter for identifying the rheological properties of silver conductive inks. The viscosity values of the PEG/glycerol/ DI water mixture are presented in **Figure 2**. From the figure, it is evident that all the fractions in the pink area have viscosity values in the range of 10 to 30 cps, indicating that all the matrix formed from these fractions are compatible with inkjet printing. These results were confirmed by similar literature works [12,13]. This ensures that the conductive ink will flow through the nozzle smoothly without experiencing problems related to leaking, drying, or clotting. On the other hand, we validated that the dispersion of filler in the suspension matrix does not have a great influence on the viscosity. Based on these results, we choose to study the viscosity of the suspension matrix as a major parameter. Different Formulations were prepared with a variable amount of glycerol 0, 20, 40, and 50%wt and a fixed amount of AgNPs (2.5 %wt) to study the effect of glycerol on the conductive properties of developed ink.

3.3. SEM analysis

The microstructure of the conductive lines printed on textile support was characterized to determine the correlation between the formulation and ink properties. **Figure 3** shows SEM images of the printed lines using the four different formulations. Rough, irregular shapes and the presence of small holes can be easily identified from the printed conductive lines. This is because the presence of a solvent with a high evaporation temperature (Glycerol 290 °C) tends to produce small bubbles during drying, leaving more voids in the printed lines as shown in **Figure 3-C, D**[11]. For formulations with 0% and 20% glycerol, the silver nanoparticles form an almost compact network in the printed lines after drying, which allows for efficient conductivity with minimal risk of short circuits and energy loss. In addition, for the formulation with 0% glycerol, although the printed line has a clear edge and its surface is slightly wrinkled and less uniform compared to that printed with 20% glycerol.

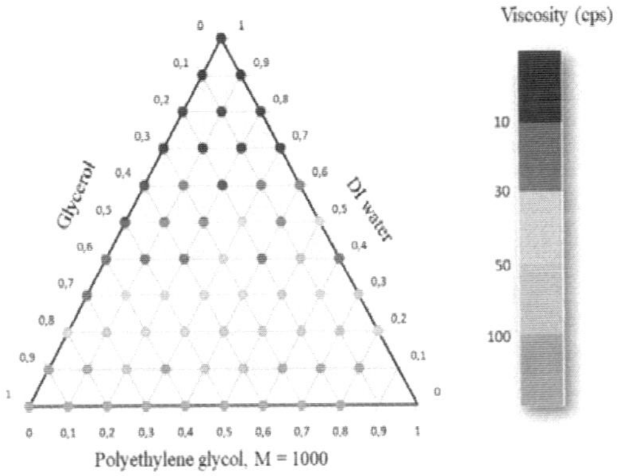

Figure 2: Viscosity of the mixture of glycerol, PEG, and DI water.

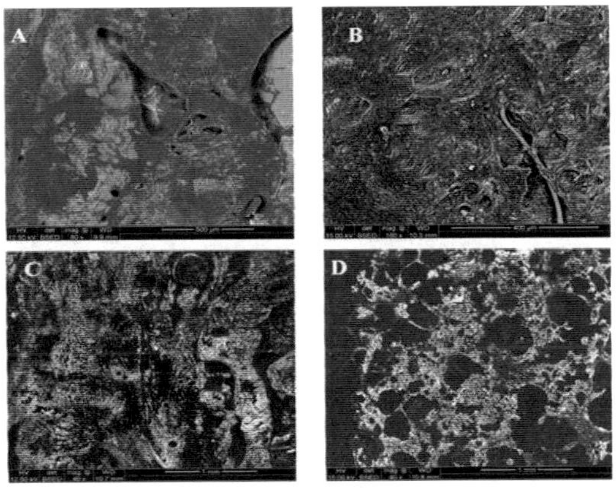

Figure 3: SEM images of conductive printed lines with 2,5 %wt of AgNPs and different glycerol content: A) 0%wt, B) 20%wt, C) 40%wt, and D) 50%wt.

3.4. Electrical measurement.

To study the electrical properties of the conductive printed lines the resistance measurements were performed using the two rods method. As shown in **Figure 4**, it can be seen that the variation of the resistance of the conductive lines is proportional to the weight percentage of glycerol. This is because the conductivity of the printed line can be influenced by the formulation of the suspension matrix [14]. The resistance of the printed line decreases with the decrease of the percentage of glycerol until a minimum value corresponds to its dispersion (0%wt). The recorded resistance value is 0.82MΩ\cm, which allows the printed textile to be oriented toward flexible electronic applications. These results show that the formulation of ink only with PEG can lead to a stable formulation with optimal conductivity. This is true if considering the properties of PEG as dispersants and capping agents.

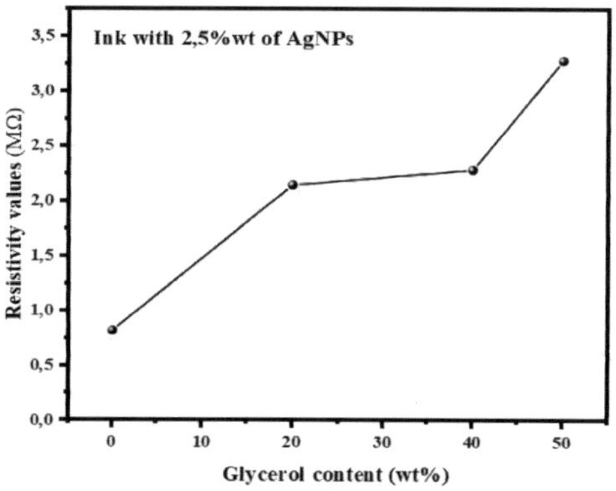

Figure 4: Electrical resistivity of the printed lines using different glycerol content.

4. Conclusion

In summary, the dispersion of silver nanoparticles in a PEG/Glycerol/DI water mixture has been developed for highly conductive wearable E-textile applications. The viscosity of the suspension matrix was studied as a critical property parameter for identifying the rheological properties of silver conductive inks. The printed textile exhibited a resistance of the conductive lines which is proportional to the weight percentage of glycerol. The conductive ink pattern applied to the cotton cloth has a conductivity value of 0.82MΩ\cm, in absence of glycerol. There are two ways to improve the conductivity of silver ink and textile substrate. One is to consider the effect of thermal sintering and increase the silver time [15]. The other is to look for other elements that affect the improvement of the suspension matrix without affecting the conductivity of the nanoparticles. Our future work will focus on the development of a suspension matrix to obtain a higher conductivity and to be applicable in the electronics field.

Acknowledgment:

The authors gratefully acknowledge the laboratory of Research on Textile Materials (REMTEX) from the Higher School of Textile and Clothing Industries (ESITH) as well as the Ain Chock Faculty of Sciences for their financial support and technical assistance. In addition, Royal Gendarmerie Forensic Science Institute, for its technical contribution.

References

[1] Eggers H, Schackmar F, Abzieher T, Sun Q, Lemmer U, Vaynzof Y, Richards B S, Hernandez-Sosa G and Paetzold U W 2020 Inkjet-Printed Micrometer-Thick Perovskite Solar Cells with Large Columnar Grains *Advanced Energy Materials* **10** 1903184

[2] Kim S 2020 Inkjet-Printed Electronics on Paper for RF Identification (RFID) and Sensing *Electronics* **9** 1636

[3] Sowade E, Mitra K Y, Ramon E, Martinez-Domingo C, Villani F, Loffredo F, Gomes H L and Baumann R R 2016 Up-scaling of the manufacturing of all-inkjet-printed organic thin-film transistors: Device performance and manufacturing yield of transistor arrays *Organic Electronics* **30** 237–46

[4] Kim N W, Lee D-G, Kim K-S and Hur S 2020 Effects of Curing Temperature on Bending Durability of Inkjet-Printed Flexible Silver Electrode *Nanomaterials* **10** 2463

[5] Gozutok Z, Agırbas O, Bahtiyari M I and Ozdemir A T 2021 Low-voltage textile-based wearable heater systems fabricated by printing reactive silver inks *Sensors and Actuators A: Physical* **322** 112610

[6] Beduk T, Bihar E, Surya S G, Castillo A N, Inal S and Salama K N 2020 A paper-based inkjet-printed PEDOT:PSS/ZnO sol-gel hydrazine sensor *Sensors and Actuators B: Chemical* **306** 127539

[7] Kholghi Eshkalak S, Chinnappan A, Jayathilaka W A D M, Khatibzadeh M, Kowsari E and Ramakrishna S 2017 A review on inkjet printing of CNT composites for smart applications *Applied Materials Today* **9** 372–86

[8] Stempien Z, Rybicki E, Rybicki T and Kozanecki M 2016 Reactive inkjet printing of PEDOT electroconductive layers on textile surfaces *Synthetic Metals* **217** 276–87

[9] Nie X, Wang H and Zou J 2012 Inkjet printing of silver citrate conductive ink on PET substrate *Applied Surface Science* **261** 554–60

[10] Boumeganane A, Nadi A, Cherkaoui O and Tahiri M 2022 Inkjet printing of silver conductive ink on textiles for wearable electronic applications *Materials Today: Proceedings* **58** 1235–41

[11] Hong H, Jiyong H, Moon K-S, Yan X and Wong C 2021 Rheological properties and screen printability of UV curable conductive ink for flexible and washable E-textiles *Journal of Materials Science & Technology* **67** 145–55

[12] Xiong Z and Liu C 2012 Optimization of inkjet printed PEDOT:PSS thin films through annealing processes *Organic Electronics* **13** 1532–40

[13] Kao H, Chuang C-H, Chang L-C, Cho C-L and Chiu H-C 2019 Inkjet-printed silver films on textiles for wearable electronics applications *Surface and Coatings Technology* **362** 328–32

[14] Denneulin A, Bras J, Carcone F, Neuman C and Blayo A 2011 Impact of ink formulation on carbon nanotube network organization within inkjet printed conductive films *Carbon* **49** 2603–14

[15] Ryu K, Moon Y-J, Park K, Hwang J-Y and Moon S-J 2016 Electrical Property and Surface Morphology of Silver Nanoparticles After Thermal Sintering *J. Electron. Mater.* **45** 312–21

IOP Conf. Series: Materials Science and Engineering 1266 (2023) 012007 doi:10.1088/1757-899X/1266/1/012007

Smart apparel design for urinary incontinence detection

A Coulombe[1], J Guérineau[1] and N Mezghani[2]

[1] Vestechpro, Cégep Marie-Victorin, 7000 Rue Marie-Victorin, Montréal, QC, H1G 2J6, Canada
[2] Institut d'intelligence artificielle appliquée, Teluq University, Montréal, QC, H2S 3L4, Canada

julia.guerineau@vestechpro.com

Abstract. A wide range of wearable devices are now used to help people with various health conditions. While approximately 10% of the Canadian population is affected by some form of urinary incontinence, there is a significant need for devices addressing this condition. This paper presents an ongoing research project for the design and development of an underwear fitted with urinary detection capacities. The paper focuses on the testing and comparison of three different solutions identified from scientific literature for detecting urinary leakage, namely by measuring conductivity, temperature, and humidity. These three detection modules have been integrated into a single prototype to ensure that they are tested under the same conditions. Our results point out that conductivity and humidity measurements appear to be viable for urine leakage detection in an absorbent pad, whereas temperature measurement has proven to be unsuccessful due to the rapid drop of the solution temperature and the time required for the liquid to reach the sensor. The temperature method is hence excluded from the next development stages. Finally, further tests on participants are still required to evaluate how body fluids other than urine might impact conductivity and humidity measurements.

1. Introduction

Recent developments in wearable electronics are now improving the quality of healthcare as well as the quality of life of many patients. Indeed, an increasing number of health practitioners are using them to monitor the health status of their patients in a continuous and remote way. Wearable devices also allow for the collection of a greater amount of data as they are typically meant to be worn for a longer period. For certain medical conditions, this data can be used for algorithm development to predict and detect health events [1]. This is notably the case for urinary incontinence (UI), a stigmatizing medical and social issue defined as any "complaint of involuntary loss of urine"[2]. In Canada, approximately 3.3 million people are affected by some form of UI [3].

Although there are a few connected devices for incontinence on the market, they are generally designed to be worn by newborns or children. As for those intended for adults, most of them are used by health professionals within a healthcare facility. Others are designed to be placed on the bed and are therefore only useful when the patient is lying down. Most importantly, to our knowledge, none of them foster patient dignity and autonomy.

A wearable fitted with both detection and prevention capabilities has been conceptualized to meet these needs. Since the detection module will be used for the development of prevention algorithms, the first step is to ensure that it accurately detects urinary leakages. Hence, this article focuses on the ongoing development of an absorbent pad integrating a detection module for autonomous individuals with functional incontinence [4] or with a reduced or absent bladder filling sensation.

Three methods for urinary leakage detection were identified and selected from the literature for their estimated cost, their wearability, and their technology readiness level: (1) conductivity [5–7], (2) temperature [7–10] and (3) humidity [11,12]. Although each of these three detection methods is well documented in the literature, to our knowledge, there is no study comparing their accuracy, advantages and limits when integrated in an absorbent pad. Accordingly, this paper aims at identifying which of humidity, temperature and conductivity measurements are the most suitable and accurate for detecting urine leakage in an absorbent pad.

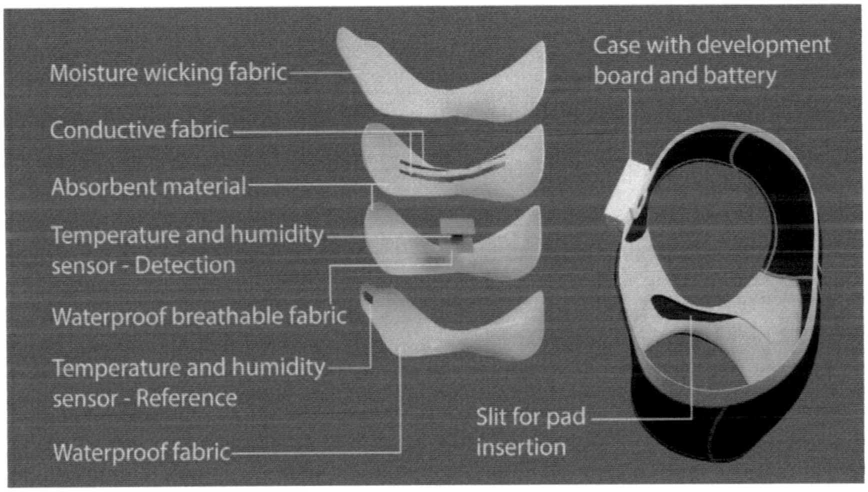

Figure 1. The prototype and its core components.

2. Material and methods

As depicted in Figure 1, the envisioned wearable takes the form of a regular underwear with an insertable absorbent pad. For testing purposes, the three detection modules have been integrated within a single prototype to ensure that they are all tested under the same conditions. The tests were performed on the absorbent pad laid flat.

For detection with conductivity, two parallel pieces of conductive fabric were placed in the absorbent pad and connected to the microcontroller (SparkFun ESP32 Thing) in an open circuit. When urine is absorbed by the pad, it creates an electrical connection between the two pieces of conductive fabric and thus closes the circuit (see Figure 2). Once the circuit is closed, a signal is sent by the microcontroller to indicate a urine leakage.

For detection based on temperature and humidity, the same sensor was used, the AHT20[3] since it can measure both. For both measurements, a detection sensor was integrated within the absorbent material, nearby the urethra, and a reference sensor was added in the pubic region for comparison (see Figure 3).

According to Eckford *et al.*, urine temperature may vary, but is always higher than 35.7 °C [8]. The microcontroller was hence programmed to detect a urine leakage with temperature when two conditions are met: (1) the detection sensor measures a temperature higher than 35.7 °C and (2) the temperature of the detection sensor is higher than the one measured by the reference sensor. Regarding detection with humidity, the code was written so that the microcontroller detects a urine leakage when (1) the relative humidity measured by the detection sensor is greater than or equal to 95% and (2) the one measured by the reference sensor is lower than 95%.

[3] https://www.adafruit.com/product/4566

IOP Conf. Series: Materials Science and Engineering 1266 (2023) 012007 doi:10.1088/1757-899X/1266/1/012007

Since the AHT20 sensor can be damaged by liquids, different waterproof and breathable fabrics have been tested to encapsulate it without affecting temperature and humidity measurements. These tests involved integrating a AHT20 sensor in a fabric pouch sealed with ultrasound and then immersing the pouch in warm water. The final choice of fabric to encapsulate the temperature and humidity sensor was based on four criteria, namely (1) its breathability, (2) how waterproof it is, (3) its thermal conductivity and (4) if it is well suited for ultrasound sealing. The chosen fabric is the 3L-WPB-STRETCH from Rockywoods.[4]

To simulate the average perineal temperature of a person sitting with legs uncrossed, a heating mat set to 34.1 °C was used [8] and folded over the prototype. A saline solution composed of 97.2% of water and 2.8% of sodium chloride at a temperature of 35.8 °C was used to imitate urine. To simulate a urine leakage, a syringe was used to inject 70 mL of this solution onto the absorbent pad, where the urethra would be located. Data were collected for 5 minutes after the injection and the pad had to be completely dry prior to the next test.

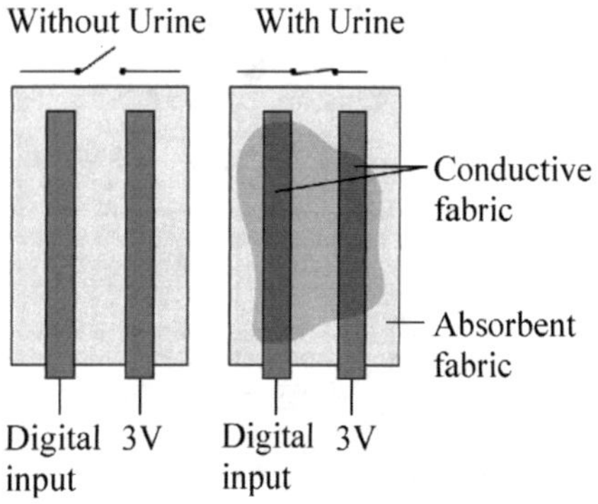

Figure 2. Basic principle for detecting urinary leakage with conductivity.

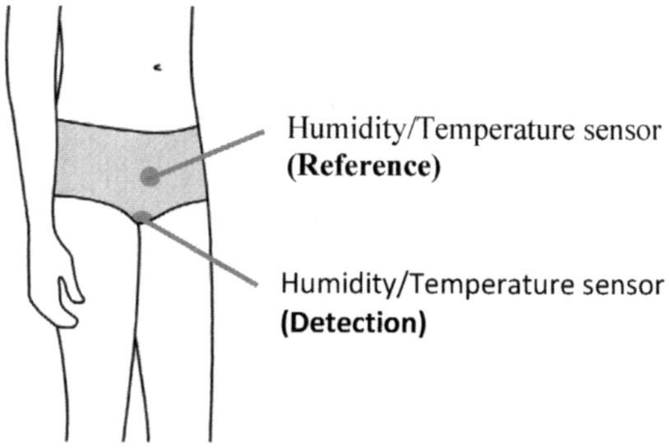

Figure 3. Basic principle for detecting urinary leakage with temperature or humidity.

3. Results and discussion

Ten tests were performed using the same prototype. The results are gathered in Table 1. For each test, the detection result is indicated (Yes/No), as well as the timespan between injection and detection – if a detection occurred. From an overall perspective, the simulated leakage was detected in ten out of ten tests with both methods using conductivity and humidity measurements. On the contrary, detection with temperature only worked in one out of ten tests.

With an average detection time of 19 seconds, measuring conductivity appears to be the fastest method for detecting urinary leakage in an absorbent pad among the three methods tested. In terms of advantages, the conductivity module has a great wearability. Indeed, it can be easily made with soft electronics, such as conductive fabric, conductive embroidery, or conductive ink, and is therefore well suited for a device intended to be thin and subtle. However, detection with conductivity does not allow to detect consecutive leakages as the pad needs to dry before detecting a new one. In addition, tests on participants are still required to determine how other body fluids, such as sweat or blood, might impact the results and whether repetitive contact with urine can degrade the conductive elements.

Similarly, other body fluids may interfere with humidity measurements, and would require further tests on participants. Furthermore, the prototype must also dry sufficiently before being able to detect a new leakage with humidity. Nonetheless, humidity measurements proved to be an avenue to further investigate.

Based on Cusick *et al.*, it was expected that temperature measurement would allow for the detection of multiple leaks without requiring the prototype to dry out, as long as temperature had time to lower

[4] https://www.rockywoods.com/3-Layer-Waterproof-Breathable-Polyester-Fabric-with-Stretch-Wild-Lime

between events [13]. However, this could not be validated since leakage was only detected in one test. Detection based on temperature appears to be unreliable since the saline solution temperature drops rapidly after injection and leaves very little time for it to be detected by the temperature sensor. In fact, detection with temperature failed in nine out of ten tests, because the saline solution temperature was already below 35.7 °C by the time it was absorbed by the matter surrounding the sensor and passed through its protecting pouch. Nonetheless, it is important to note that the low detection rate can be due to the prototype construction and that it may be possible to overcome this problem by positioning the sensor closer to the pad surface or by using materials that channel the liquid towards the sensor. Regardless of how the prototype is built, other issues are expected to arise during tests on participants. Indeed, according to data collected by Eckford *et al.*, the perineal temperature can sometimes be higher than the temperature of urine itself, especially when a person is sitting with legs crossed [8]. This temperature increase in the perineal region can lead to false positive results. Therefore, according to our test results, measuring temperature with a fabric-encapsulated sensor placed between two layers of absorbent matter is not a reliable option as the sole means of detecting urine leakage.

	Detection with conductivity		Detection with temperature		Detection with humidity	
Test #	Yes/No	Detection time (sec.)	Yes/No	Detection time (sec.)	Yes/No	Detection time (sec.)
1	Yes	22	No	-	Yes	87
2	Yes	18	No	-	Yes	48
3	Yes	16	Yes	42	Yes	88
4	Yes	13	No	-	Yes	58
5	Yes	14	No	-	Yes	49
6	Yes	17	No	-	Yes	57
7	Yes	15	No	-	Yes	48
8	Yes	28	No	-	Yes	69
9	Yes	18	No	-	Yes	60
10	Yes	24	No	-	Yes	36

The prototype development will thus continue with detection methods using conductivity and humidity measurements. Both solutions have proven to be individually viable and their possible complementarity will be investigated during further tests on participants.

4. Conclusion

This paper focuses on identifying which of humidity, temperature and conductivity measurements are the most suitable and accurate for detecting urine leakage in an absorbent pad. The conducted tests have shown that conductivity and humidity methods appear to be viable candidates for the detection of urinary incontinence events and should be further investigated with tests on participants. A combination of these two detection methods is also considered. However, solely relying on temperature measurements to detect urine leakage has proven to be challenging as urine temperature drops rapidly after voiding and since perineal temperature can sometimes be higher than urine temperature [8]. For future work, a complementary module for urine leakage prevention will be developed using ultrasounds and/or body impedance along with algorithms for bladder volume assessment.

Acknowledgments
This work has been funded by Quebec's "Fonds de Recherche Nature et technologies – Programme Audace". The authors are grateful to all the consortium partners for their valuable contributions during the development of ideas and concepts.

Code availability
The Arduino code used to perform the tests is available at [14].

References

[1] Dunn J, Runge R and Snyder M 2018 Wearables and the medical revolution *Personalized Medicine* **15** 429–48

[2] D'Ancona C, Haylen B, Oelke M, Abranches-Monteiro L, Arnold E, Goldman H, Hamid R, Homma Y, Marcelissen T, Rademakers K, Schizas A, Singla A, Soto I, Tse V, de Wachter S and Herschorn S 2019 The International Continence Society (ICS) report on the terminology for adult male lower urinary tract and pelvic floor symptoms and dysfunction *Neurourology and Urodynamics* **38** 433–77

[3] The Canadian Continence Foundation 2007 *Incontinence: une perspective canadienne*

[4] Cheater F M and Castleden C M 2000 Epidemiology and classification of urinary incontinence *Best Practice & Research Clinical Obstetrics & Gynaecology* **14** 183–205

[5] Fernandes B, Gaydecki P, Jowitt F and van den Heuvel E 2011 Urinary Incontinence: A Vibration Alert System for Detecting Pad Overflow *Assistive Technology* **23** 218–24

[6] Fischer M, Renzler M and Ussmueller T 2019 Development of a Smart Bed Insert for Detection of Incontinence and Occupation in Elder Care *IEEE Access* **7** 118498–508

[7] Tamura T, Nakajima K, Matsushita T, Fujimoto T, Shimooki S and Nakano T 1993 A monitoring and telemetering system of urinary incontinence for home health care *Proceedings of the 15th Annual International Conference of the IEEE Engineering in Medicine and Biology Societ* (IEEE) pp 998–9

[8] Eckford S D, Finney R, Jackson S R and Abrams P 1996 Detection of urinary incontinence during ambulatory monitoring of bladder function by a temperature-sensitive device *British Journal of Urology* **77** 194–7

[9] Bosboom-van der Hurk P R, Middelkoop H A M, van Waalwijk-van Doorn E S C, Roos R A C and Cools H J M 1998 Long-term ambulatory monitoring of urine leakage in the elderly: An evaluation of the validity and clinical applicability of thermistor signalling *Journal of Medical Engineering & Technology* **22** 91–3

[10] Ejaz T, Nakae T, Takemae T, Egami C, Sugihara O and Ikeda H 1998 A sensing system for simultaneous detection of urine and its components *IEEE. APCCAS 1998. 1998 IEEE Asia-Pacific Conference on Circuits and Systems. Microelectronics and Integrating Systems. Proceedings (Cat. No.98EX242)* (IEEE) pp 221–4

[11] Shri Harini G, Vishal N and Prince Sahaya Brighty S 2020 Incontinence Monitoring System Using Wireless Sensor for "Smart Diapers" pp 1324–31

[12] Ang L M, Ow S H, Seng K P, Tee Z H, Lee B W, Thong M K, Poi P J H and Kunanayagam S 2008 Wireless intelligent incontinence management system using smart diapers *2008 5th International Conference on Electrical Engineering/Electronics, Computer, Telecommunications and Information Technology* (IEEE) pp 69–72

[13] Cusick G, Birkett A, Clarke-O'Neill S, Fader M and Cottenden A M 2003 A system for logging incontinence events using a simple disposable sensor *Proceedings of the Institution of Mechanical Engineers, Part H: Journal of Engineering in Medicine* **217** 305–10

[14] Coulombe A, Guérineau J and Mezghani N 2022 Smart apparel design for urinary incontinence detection: Arduino code *Zenodo*

Simulation techniques for smart textile predictive design

Caitlin G. Knowles[1], Beomjun Ju[1], Busra Sennik[1], Amanda C. Mills[1], Jesse S. Jur[1]

[1]Department of Textile Engineering, Chemistry and Science, North Carolina State University, USA

Corresponding author: jsjur@ncsu.edu

Abstract. Smart textiles must combine both textile and electronic systems into one product. This presents challenges as each industry has their own design and evaluation standards that are not compatible with one another. As such, smart textile designers tend to rely heavily on the production and iteration of physical prototypes to create a product that meets the specified design criteria. One emerging tool in the apparel industry that has potential to shorten the prototyping cycle is 3D CAD for textiles, also known as 3D garment simulation. While typically used for apparel design and e-commerce, this work presents two case studies that demonstrate how 3D garment simulation can be used as a tool for predictive design of smart textile products. In particular, how strain-dependent properties such as resistance and contact pressure can be predicted and how designs can be optimized to achieve certain performance metrics.

1. Introduction

Smart textiles are faced with the fundamental challenge of combining textile and electronic systems into one product. The textile and electronics industries have different design strategies and performance metrics which can make producing prototypes challenging. The electronics industry has a well-defined set of digital and automated tools to design, simulate, and manufacture products. In comparison, the textile industry has a predominantly physical design cycle, relying on physical prototypes and manual manufacturing processes (e.g., cut and sew). Recently, 3D CAD tools for apparel have emerged, driven by companies including CLO Virtual Fashion, Browzwear, Lectra, and Optitex. These 3D design tools such as particle-based 3D garment simulation are becoming widely adopted for reasons such as improved quality, flexible manufacturing, reduced time to market, reduced cost, and sustainability [1]–[3].

While increasingly common within the apparel industry for these purposes, 3D garment simulation is less often used as a tool to predict performance and fit. This is due to many reasons, most notably the issues with simulation accuracy such as the oversimplification of a textile's non-linear, viscoelastic mechanical properties to a more simply modeled linear stress response [4]. This highlights a tradeoff between computational time and accuracy for particle-based cloth simulation and more trusted methods such as finite element analysis.

This paper describes techniques using 3D garment simulation to predict the performance of smart textile systems. Two case study examples are provided: first, the prediction of screen-printed interconnects' electromechanical performance in the form of relative resistance increase with elongation ($\Delta R/R_0$) and second, the contact pressure and data quality (signal to noise ratio, SNR) prediction of a biopotential monitoring armband. In addition to proposing simulation techniques, this work aims to fill two gaps in smart textiles literature. First, while several studies evaluate how the geometry of the interconnect impacts the electromechanical performance [5], [6], only limited studies investigate the impact of the substrate [7]. Second, for biopotential armbands, several authors have investigated the impact of contact pressure on the ECG data quality [8], [9] but no researchers to the authors' knowledge have proposed accessible design strategies to achieve the desired contact pressure based on the arm geometry and fabric properties.

2. Materials and Methods

The smart textile components and systems were made via screen-printing on thermoplastic polyurethane (Dupont TE-11C) with silver ink. A 280-mesh frame was used to print the silver ink (Dupont PE 874 and 5880). The interconnects and electrode connection points were encapsulated with Dupont TE-21C. The components (interconnects and electrodes) were laser cut to shape (Epilog Mini 18, 40 Watt) and adhered to the polyester-spandex knit textiles via heat press (Digital Knight DK20S) at 130°C for 2 minutes. Snap connections (Dritz, Size 16) were added to the electrodes on the armbands. The armbands were then sewn using a straight stitch and a 1 cm seam allowance.

The interconnects underwent electromechanical testing using a universal testing machine (Criterion Model 43, MTS) with a gauge length of 10 cm, a testing rate of 1 mm/s, and a sample acquisition rate of 10 Hz. The electrical resistance was measured using a multimeter (model 3441A, Keysight) and a data logger (Keysight BenchVue). For the biopotential armbands, the electrocardiogram (ECG) signal was measured using a BioPac MP16 system interface with an ECG100C wired amplifier attachment. A 100 Hz sampling frequency was used, and the signal was captured for 1-minute, following a 2-minute settling period.

The single jersey polyester-spandex knit fabrics (Spandex by Yard) had variable spandex compositions (0-18%) and areal densities (150-350 gm^{-2}). The naming convention used for the fabrics within this paper is SA BBB, where A = the spandex content, and BBB = the areal density. Following this convention, S0 150 is 100% polyester with an areal density of 150 gm^{-2}.

Simulations for both the interconnects and armbands were created using CLO3D (CLO Virtual Fashion). The physical properties (stretch, bending, weight, thickness) of the polyester-spandex fabrics and TPU films were measured using the CLO Fabric Kit 2.0. To apply strain to the interconnects, an inextensible fabric was wrapped around a frictionless box. The level of applied strain was adjusted by changing the length of the inextensible fabric. For the armband, a Light Detection and Ranging (LiDAR) scan of the subject's arm was taken using an iPad Pro. This was then imported to an open-source 3D computer graphics software (Blender) to scale and crop before the OBJ file was imported to CLO3D as the avatar for simulation.

3. Results and Discussion

Two case studies are presented to show examples of how 3D garment simulation can be used as a tool for smart textile predictive design.

3.1. Interconnect Case Study

The interconnect case study aimed to explore how fabric selection impacts the electromechanical performance of a screen-printed interconnect. The results of the study are shown in Figure 1. Figure 1a and 1b show the experimental and simulated results for interconnects with four different polyester-spandex knit fabric substrates of varying spandex content and areal density. The images presented are at an applied strain level of 25%. In Figure 1c, strain and stress maps are shown for the S4 160 fabric substrate at applied strains from 5-35%. These maps show the stress concentration within the interconnect as well as how the surrounding substrate serves to relieve strain from the interconnect as a result of its lower modulus. To apply strain to the interconnect within the simulation environment, the 10 cm fabric was attached to an inextensible band and wrapped around a rectangular box. The level of applied strain was adjusted by changing the length of the band.

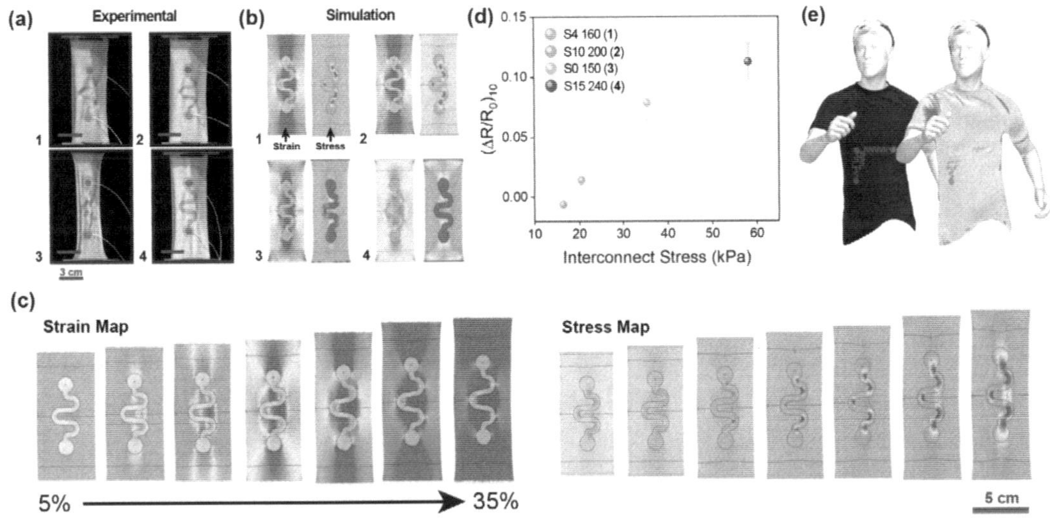

Figure 1. Interconnect simulation and electromechanical analysis.

As these results show, the fabric substrate and its stiffness have a substantial impact on the strain and stress distribution within the interconnect system. Within the simulation at an applied strain of 25%, the localized strain in the fabric surrounding the interconnect varied from 29%-54%. This represents the strain relieving nature of the fabric and resulted in simulated localized strains within the interconnect that ranged from 6%-15% and stresses ranging from 56 kPa-153 kPa.

The substrate impact on localized interconnect strain and stress is also present in the experimental electromechanical analysis. Figure 1d shows the simulated interconnect stress at 10% applied strain plotted against the relative resistance change ($\Delta R/R_0$) at the same strain level. The positive correlation is strong, with r=0.94. A correlation analysis between the fabric material properties and the resulting performance showed that the stiffness in the stretch (weft) direction was closely related to the resistance increase. Fabrics S0 150 and S15 240 have high stiffness (E_{Weft} = 226 kPa, 256 kPa), causing the interconnect to undergo greater strain relative to the surrounding substrate and experience a greater resistance increase. In contrast, fabrics S10 200 and S4 160 have low stiffness (E_{weft} = 134 kPa, 256 kPa) and accommodated more of the system strain, causing the interconnect to stretch less and have a lower increase in resistance.

Understanding that the simulated stress from CLO3D is correlated with the electromechanical performance allows for the extension of this method to full-garment designs (Figure 1e). One of the challenges in e-textile design is determining the path of the electronics across the human body and the impact of the fabric selection on the performance of the overall system is often overlooked. The use of 3D garment simulation at this stage is a simple and accessible technique to help determine the routing of the interconnects as well as the fabric selection to minimize resistance increase.

3.2. Biopotential Armband Case Study

The biopotential armband case study was designed to address the challenge of designing a garment to achieve certain contact pressure on the body. For biopotential electrodes, a certain contact pressure must be reached to obtain low skin-electrode impedance and minimize motion artifacts. 3D garment simulation is proposed as a tool to help inform the designer in achieving the desired contact pressure and resulting data quality.

The results of the biopotential armband case study are shown in Figure 2. The electrode design is shown in Figure 2a with the printed Ag/AgCl electrode on a TPU substrate with a 2 mm offset. The snap

IOP Conf. Series: Materials Science and Engineering 1266 (2023) 012008 doi:10.1088/1757-899X/1266/1/012008

connection allows for the collection of the electrocardiogram (ECG) signal from each of the three electrodes.

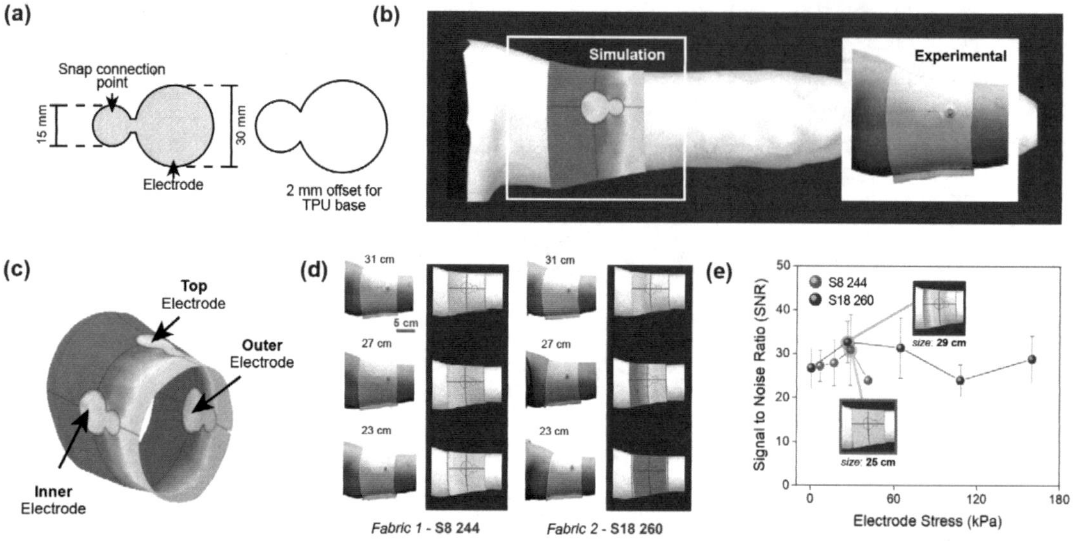

Figure 2. Biopotential armband simulation and experimental results

Figure 2b shows the simulated armband within CLO3D on the LiDAR-scanned arm of the subject as well as the experimental comparison. The three dry Ag/AgCl electrodes are located at the top, inner, and outer portions of the subject's upper arm (Figure 2c).

One of the limitations of 3D garment simulation is that the avatar is rigid and thus the simulation is unable to capture the soft tissue deformation that is present with compressive garments. This impact is captured in the experimental images and their simulated counterparts shown in Figure 2d. A subset of 3 of the 5 sizes tested are shown, ranging from 23 cm to 31 cm. The two fabrics differ substantially in stiffness (E = 84 kPa and 239 kPa respectively, for S8 244 and S18 260) and thus result in different levels of simulated fabric stress and experimentally measured contact pressure. Additionally, as shown in Figure 2d, the stress varied substantially along the length of the arm as the arm circumference varied from 29 cm to 33.5 cm. This meant that accurate placement was essential to achieve repeatable measurements. An additional limitation illustrated in Figure 2d is that at small sizes (large strains), the fabrics were prone to curling and rolling at the upper portion of the armband. This curling is not present in the simulation as it does not consider edge effects and will thus impact the accuracy of the result.

For each armband size and fabric, the ECG signal was measured on a male subject and the results were analyzed to extract the signal to noise ratio (SNR). The simulated stress values were then plotted against the biopotential data quality in the form of SNR in Figure 2e. This plot shows that fabrics will achieve the same stress level at different sizes based on their physical properties. S8 244 and S18 260 both peak around the same electrode stress value (30 kPa), although the sizes are different (25 cm vs. 29 cm). For S8 244, the SNR peaked with a SNR of 30.9 with a large sample-to-sample standard deviation of 26.1%. S18 260 peaked with a SNR of 32.7 and standard deviation of 14.7%. Further research is necessary including more accurate contact pressure measurement and dynamic ECG measurement to better capture the impact of motion artifacts. However, these preliminary results suggest that 3D garment simulation may be used to adjust size and fabric selection to achieve the desired ECG data quality.

4. Conclusion

The results presented here in the form of two case studies show the potential of 3D garment simulation as a tool to predict smart textile performance, either in the form of the relative resistance increase, or the contact pressure and signal to noise ratio (SNR). The proposed technique is accessible by engineers and designers alike and requires only basic testing equipment. While limitations in simulation accuracy exist and therefore direct predictions cannot be made, the results can help narrow a wide selection of designs and fabrics and thus shorten the prototyping process. Considerable opportunities exist to expand upon this research in the field of smart textiles. Strain-dependent properties (such as resistance and contact pressure) are most promising to predict as the simulation outputs are currently limited to stress and strain maps.

Acknowledgment: The authors would like to express appreciation for the support of the Digital Education and Learning Technology Applications (DELTA) at North Carolina State University as well as Advanced Functional Fabrics of America (AFFOA).

References

[1] E. Lee and H. Park, *Int. J. Fash. Des. Technol. Educ.*, vol. 10, no. 1, pp. 59–70, Jan. 2017
[2] E. Ju and M. G. Choi, *IEEE Access*, vol. 8, pp. 195113–195121, 2020
[3] H. K. Song and S. P. Ashdown, *Cloth. Text. Res. J.*, vol. 33, no. 4, pp. 314–330, Oct. 2015
[4] S. Kuijpers *et al.*, *IEEE SA Industry Connections*, p. 44, 2020.
[5] R. Xu *et al.*, *Extreme Mech. Lett.*, vol. 1, pp. 120–126, Dec. 2014
[6] M. A. Yokus, R. Foote, and J. S. Jur, *IEEE Sens. J.*, vol. 16, no. 22, pp. 7967–7976, Nov. 2016
[7] C. Ye, B. G. Stewart, and S. K. Sitaraman, in *2020 IEEE 70th Electronic Components and Technology Conference (ECTC)*, Jun. 2020, pp. 1533–1541
[8] T. Takeshita *et al.*, *Sci. Rep.*, vol. 9, no. 1, Art. no. 1, Apr. 2019
[9] B. M. Li *et al.*, *IEEE Sens. J.*, vol. 21, no. 9, pp. 11046–11060, May 2021

IOP Conf. Series: Materials Science and Engineering 1266 (2023) 012009 doi:10.1088/1757-899X/1266/1/012009

Bending Properties of Fiber Composite Dilatant Compound

Shunichi Kobayashi[1, 2], **Taiki Nitta**[2], **Takumi Kobayashi**[2]

[1] Institute for Fiber Engineering (IFES), Interdisciplinary Cluster for Cutting Edge Research (ICCER), Shinshu University, 3-15-1 Tokida Ueda Nagano, Japan. (E-mail: shukoba@shinshu-u.ac.jp)
[2] Graduate School of Science and Technology, Shinshu University, 3-15-1 Tokida Ueda Nagano, Japan.

Abstract. We have developed a viscoelastic-plastic dilatant compound, whose stiffness increases with increasing deformation rate, as a matrix material, and a "fiber composite dilatant compound" to which fibers have been added. In this study, three-point bending flexural tests were conducted to clarify the mechanical properties of the material. The results showed that the bending load increased when the fibers were oriented in the bending direction. It was also found that, under constant fiber content, lengthening the fibers and increasing the number of fibers by decreasing their diameter contributed to the increase in bending load.

1. Introduction

We have developed a fiber composite fluid, in which a dilatant fluid (Sear Thickening Fluid), having a characteristic of increasing viscosity with increasing shear rate, is used as a matrix material, and fibers are added. An anisotropic viscoelastic fin encapsulating the fiber composite fluid in a flexible urethane gel container was produced, and the dependence of its stiffness on bending rate was investigated [1]. This fin can be applied to various fields by taking advantage of the feature that the stiffness changes according to the deformation rate. However, in practical use, they require a container to seal the fluid, making them structurally complex. In related research, protective textiles in which the fiber is impregnated with dilatant fluid have also been developed, but the fluid content cannot be increased [2]. Thus, we developed a "fiber composite dilatant compound" using a viscoelastic-plastic dilatant compound as a matrix material instead of dilatant fluid. Since the viscoelastic-plastic dilatant compound is not a fluid, it does not require fins for encapsulation. For the viscoelastic-plastic dilatant compound, we selected Dow Corning 3179 Dilatant Compound [2]. This is an elastic silicone polymer that is also used in a toy called Silly Putty [3]. The mechanical properties of this material have been investigated [3-4], and its potential for sensor applications has also been studied [5].

In this study, three-point bending flexural tests were conducted to characterize the stiffness of fiber composite dilatant compounds as a mechanical property that increases with the deformation rate. This paper reports the results of the tests with varying fiber orientation and fiber specifications (length, diameter, and filling ratio).

2. Material and Methods

The matrix for the fiber composite dilatant compound was Dow Corning 3179 Dilatant Compound (henceforth referred to as compound), as described in the introduction. The fibers were nylon fibers regarding their strength balance with compound and availability.

Figure 1 shows a test piece of three-point bending flexural test. The test piece was rectangular, and its dimensions was 90 × 20 × 9 mm. The diameter and length of the fiber (L and D), and fiber filling ratio (Fr, vol%) were varied. The test piece consists of three fiber layers and four compound layers (matrix layers). The method of the fabrication is as follows. (1) Forming the four compound layers (2) Placing the fibers in a certain direction on the top surface of the three compound layers (3) Overlapping the compound layers.

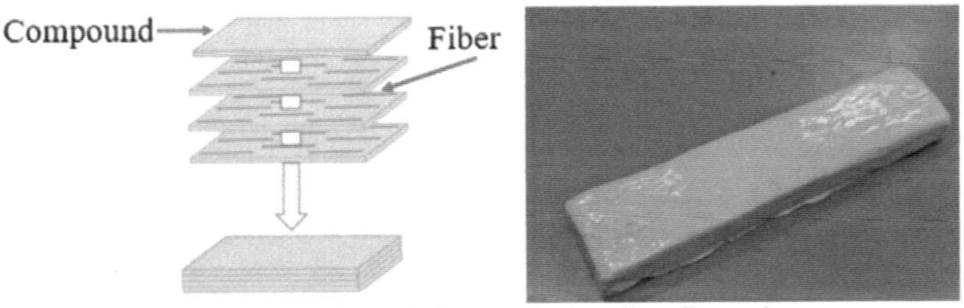

Figure 1. Test piece of three-point bending flexural test

Figure 2 shows a conceptual diagram of the effect of increased shear force obtained by composing fibers. (a) shows the shear deformation of the compound with one end of each fibers fixed. The shear deformation of the compound causes sliding between the fibers and the compound, resulting in a large shear force. (b) shows the state in which a part of the test piece is compressed and fixed, and the other tip is loaded to bent. From this figure, it can be seen that there is an area where the fiber is fixed and that the bending of the test piece creates the shear force due to sliding between the fiber and the compound. Therefore, a fixed region of fibers is necessary to generate sliding region between the fibers and the compound.

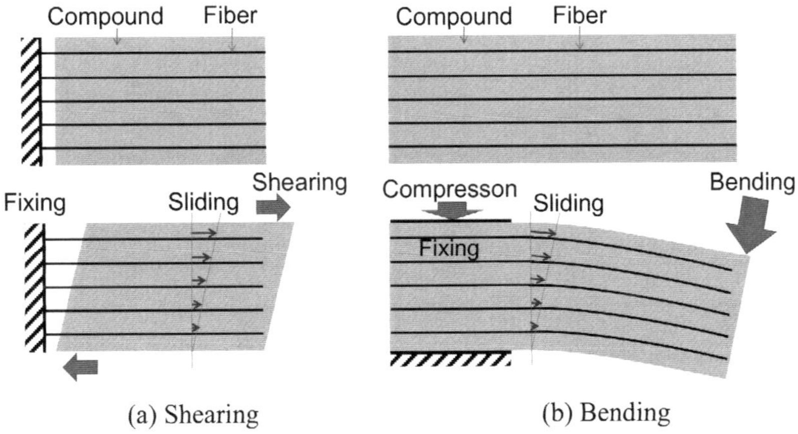

Figure 2. Concept of increasing shearing force and bending force by compositing fibers

Figure 3 shows a diagram of the three-point bending flexural test. The "parallel" fiber orientation means that the fiber direction and the direction of shear force due to bending are parallel. On the other hand, "perpendicular" means that the indenter's fiber direction and depth direction are parallel. The fiber direction and the direction of shear force due to bending are perpendicular. Indenter speed Ui was varied from 10 to 1000 mm/min from low to high speed. Indenter displacement was set to 0 mm when the indenter touched the test piece.

3. Results and discussion

3.1. Effect of fiber orientation and indenter speed on bending load

Figure 4 shows the results of a three-point bending flexural test at an indenter speed of 100 mm/min. As the displacement of the indenter increases, the bending load also increases, especially when the fiber orientation is parallel. In order to examine the effect of various test conditions on the bending load, the load at a displacement Zi of 8 mm (maximum value of the monotonically increasing load in the test) was measured three times and the average value was defined as the representative value of the bending load.

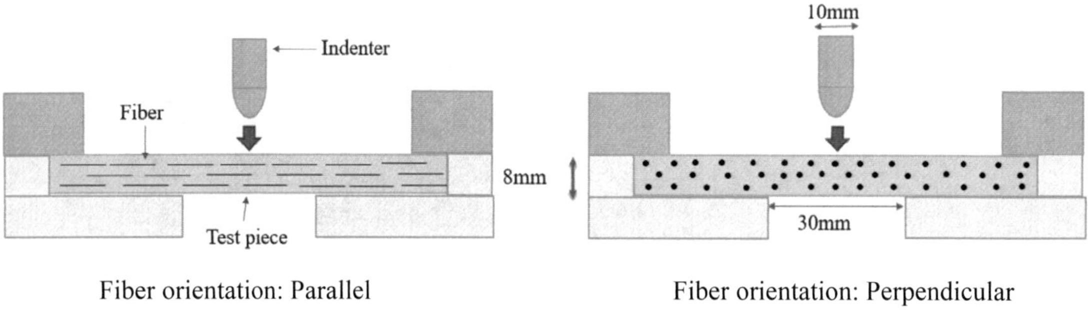

Figure 3. Three-point bending flexural test

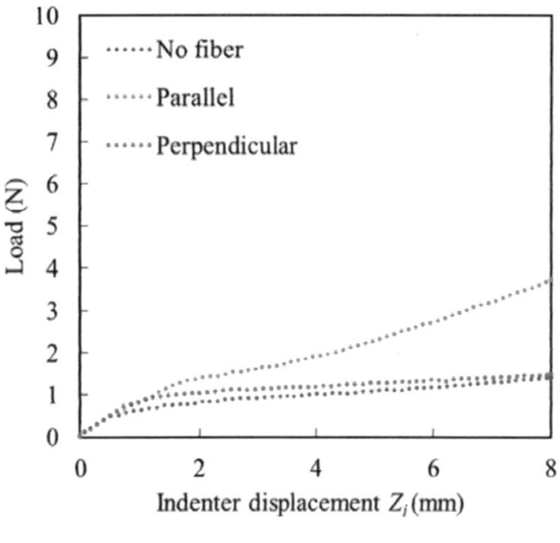

Figure 4. Three-point bending flexural test results.
Indenter speed = 100 mm/min, Fiber diameter D=0.165 mm, Fiber length, L=15mm, Number of fibers N=144, Filling ratio F_r=0.29 vol%

Figure 5 shows a schematic diagram of test piece deformation when the fiber orientation is parallel. The flow inside the compound due to the deformation produces a sliding region between the fibers and the compound. Thus, greater shear force is generated due to the shear rate of the compound. In a three-point bending flexural test, a fixation region or a region with relatively slow sliding velocity is considered to occur at the point where the compound is compressed by the indenter or support plate. As explained in Fig. 2, if there is no fixing region where the compound and fibers are fixed, a sliding region will not occur. In the three-point bending test, the fixing region is considered to occur at the point where the compound is compressed by the indenter or support plate. Thus, when the fiber orientation is parallel, a sliding region is generated and the load increases, while when the fiber orientation is perpendicular, no sliding region is generated and the load is as low as in the no-fiber condition.

Figure 6 shows the relationship between indenter speed and load at Zi=8mm. The load increases with increasing indenter speeds. These characteristics are due to the dilatancy property of the compound. The load is higher in the parallel orientation case than in the perpendicular orientation case for the

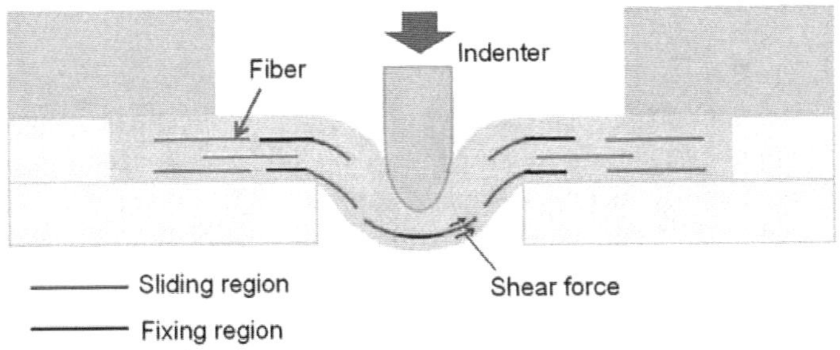

Figure 5. Schematic bending diagram of the test piece (Parallel fiber orientation)

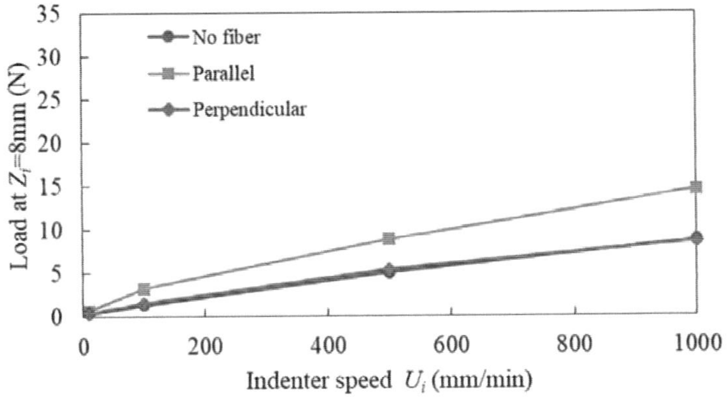

Figure 6. Relationship between indenter speed and load. Fiber diameter D=0.165 mm, Fiber length, L=15mm, Number of fibers N=144, Filling ratio F_r=0.29 vol%

fiber orientation. This is because, as already mentioned, parallel fiber orientation increases the shear force between the fibers and the compound. Also, as the indenter speed increases, it becomes even more significant due to the dilatancy of the compound.

3.2. Effect of fiber length on bending load under constant fiber diameter and fiber filling ratio (Parallel fiber orientation)

Figure 7 shows the relationship between fiber length L and load. The number of fibers varies because the fiber diameter and filling ratio are kept constant. A fiber length of 0 mm means no fiber. As L increases, the load also increases because the longer the fiber, the greater the length of the fixing and sliding regions. However, since the fiber diameter and filling ratio are kept constant, the total surface area of the fiber (excluding the area of the end faces) is constant, which means that the number of fibers is reduced when the fiber is longer. Nevertheless, the longer the fiber, the larger the load, which means that the length of the fiber is a significant factor in the load increase.

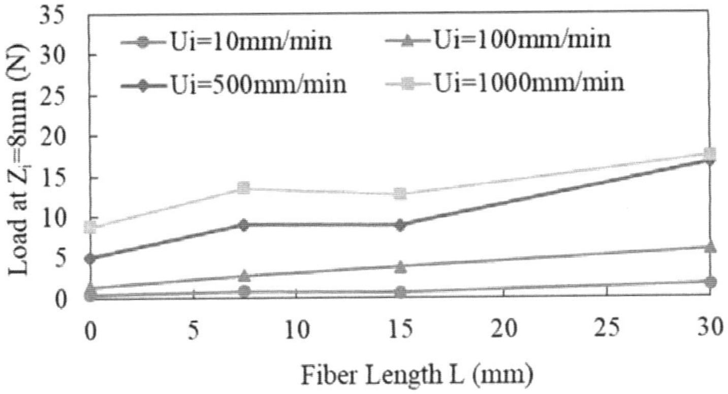

Figure 7. Relationship between fiber length L and load. Parallel orientation, Fiber diameter D=0.165 mm, Filling ratio F_r=0.29 vol %. Number of fibers N (72 - 288) were changed to keep constant Fr.

3.3. *Effect of fiber diameter on bending load under constant fiber length and fiber filling ratio (Parallel fiber orientation)*

Figure 8 shows a relationship between fiber diameter D and load. The number of fibers varies because the filling ratio and fiber length are kept constant. As in Fig. 7, a fiber length of 0 mm means no fibers, and the existence of parallel-oriented fibers increases the load. However, the load tends to decrease as the fiber diameter increases above 0.117 mm. This tendency is because since the fiber length and filling ratio is constant, the total surface area of the fibers decreases as the number of fibers decreases with increasing fiber diameter. Therefore, for the same filling ratio and fiber length, it is better to make the fiber diameter smaller to increase the number of fibers and surface area where the sliding region occurs. However, there is a limit to the smaller fiber diameter for the increase of load, and future tests with fibers with diameters smaller than 0.117 mm are needed to verify this.

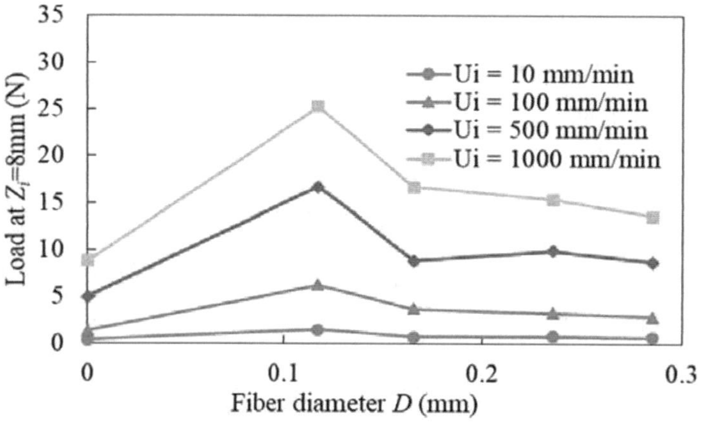

Figure 8. Relationship between fiber diameter D and load. Parallel orientation, Fiber length, L=15mm, Filling ratio F_r=0.29 vol %. Number of fibers N were changed from 288 (at D= 0.117 mm) to 48 (at D=0.285 mm) to keep constant Fr.

3.4. *Effect of filling ratio on load under constant fiber length and fiber diameter (Parallel fiber orientation)*

Figure 9 shows the relationship between fiber filling ratio Fr and load. Since the fiber length and diameter were kept constant, the filling ratio varied by changing the number of fibers. The load increased as the filling ratio increased because of the fibers' increased number and total surface area. However, the fibers' stiffness effect is negligible because this study's fiber-filling ratio is meager. From the above, the load depends on the sliding between the compound and the fibers and increasing the fibers' length, and number (surface area) is essential to increase the load.

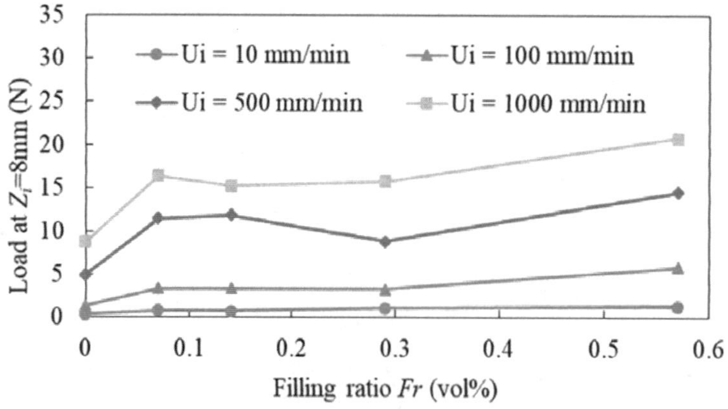

Figure 9. Relationship between fiber filling ratio Fr and load. Parallel orientation, Fiber length, L=15mm, Fiber diameter D=0.16: mm,Filling ratio F_r=0.29 vol % Number of fibers N was changed from 36 (at F_r = 0.07 vol %) to - 288 (at F_r = 0.57 vol %)

4. Conclusion

In this study, we conducted the three-point bending flexural test to investigate the mechanical properties of "The Fiber Composite Dilatant Compound." The results showed that the large bending load could obtain by orienting the fibers parallel to the bending, increasing the number of fibers by thinning and lengthening their fibers. This large bending load attributes to the increased sliding region between the fibers and the compound.

Future issues for this study include increasing the filling ratio, fiber fineness, and the effects of fiber material (fiber surface: friction between fiber and compound, and fiber stiffness).

References

[1] Kobayashi S and Sugiyama K 2019 Bending Speed Dependence Properties of Stiffness of Anisotropic Viscoelastic Fin Containing Fiber Composite Dilatant Fluid *Journal of Fiber Bioengineering and Informatics* **12** (2) pp 55-62
[2] Decker M J, Halbacha C J, Nama C H, Wagnera N J and Wetzel E D 2007 Stab Resistance of Shear Thickening Fluid (STF)-Treated Fabrics *Composites Science and Technology* **67** pp 565–578
[3] Dow Chemical Company 2017 Technical Data Sheet - DOWSIL™ 3179 Dilatant Compound
[4] Crossa R 2012 Elastic and viscous properties of Silly Putty *Am. J. Phys.* **80** (10) pp 870-875
[5] Liang J and Zhang X H 2015 Rheological Properties of SP in Shock Transmission Application *J. Mater. Civ. Eng.* **27**(9) pp 04014250-1- 04014250-7
[6] Boland C S, et. al., 2016 Sensitive electromechanical sensors using viscoelastic graphene-polymer nanocomposites *Science* **354** (6317) pp 1257-1260

IOP Conf. Series: Materials Science and Engineering 1266 (2023) 012010 doi:10.1088/1757-899X/1266/1/012010

The effect of laminate sequences on tuft bridging in GFRP laminates under tensile loading

S Marzdashti[1] and S Joncas[2]

[1]PhD Student, Ecole de technologie supérieure, Montréal, Canada

[2] Professor, Ecole de technologie supérieure, Montréal, Canada

E-mail: Sobhan.esmaeili-marzdashti.1@ens.etsmtl.ca

Abstract. This conference paper reports the tuft bridging behaviour of a single tuft when used in an orthotropic or a quasi-isotropic laminate. Laminates are subjected to pull-out experiments with a full release film placed in the mid-layer of laminates. The force-separation diagrams and damaged samples obtained by tensile tests are analysed to compare two different laminate. The results show that the larger amount of resin present around the tuft in orthotropic laminate reduces by 60% the ultimate tensile strength of the tuft when compared to quasi-isotropic laminates. The tuft tends to fail in the mid-layer of quasi-isotropic laminate with no pull-out, while the tuft has a bit pulling out in orthotropic laminate, causing an increase in the fracture energy.

1. Introduction

Due to the increase in use of aeronautics and land transportation systems made of composite materials, the production of high-performance composite parts has more than doubled in the last two decades. Among all the advantages that composite materials have, they have a relatively weak resistance to delamination and impact, especially in Mode I and transverse loading conditions. The recent development of composites made with 3D textiles or through-thickness reinforcement techniques (commonly named TTR composites) has opened many options for structural design to better control out-of-plane loads.

The most common TTR techniques are z-pinning, stitching and tufting. These techniques can improve material reliability and resistance through-thickness under different loadings [1-3]. Being cost-effective and not requiring complex equipment made the tufting technique more appealing for industrial applications than other TTR techniques in the last decade [4]. In the open literature, the effect of laminate sequences on the tuft bridging of tufted laminates has not been well addressed yet [5]. The aim of this study is to determine the differences in tuft bridging behaviour between orthotropic and quasi-isotropic laminates under tensile load. Laminates are subjected to pull-out experiments with a full release film placed in the mid-layer of laminates. The force-separation diagrams and damaged samples obtained by tensile tests are analyzed to compare tuft bridging within two different laminates.

2. Materials and experiment

2.1. Manufacturing of tufted laminate

The TTR laminates containing a single tuft were tufted by CTT GROUP using a twill 2/2 woven E-glass fabrics (TG-09-T (305 g/m2)) from Texonic [6]. The preforms are 30 plies of either consecutive layers of 0/90 for the orthotropic laminate or consecutive layers of 0/90 and ±45 for the quasi-isotropic laminate. A full release film placed in the preform at mid-thichness to isolate tuft contribution during the tensile experiments. The tufts are twisted 2 × 1k-67TEX Tenax® HTA-40 carbon fiber threads which are vertically inserted using a KSL (Keilmann Sondermaschinenbau GmbH) RS 522 tufting head as shown in Figure 1. The resin system used for infusion is Huntsman's Araldite® LY 8615/Aradur® 8615 epoxy (100:50 mass ratio). The system was infused and cured at 40°C for 24h, followed by post-curing at 180°C for 3h. As shown in Figure 2, the cured laminates were cut into 8mm×8mm cross-sections samples containing a single tuft using a radial abrasive machine.

Figure 1. (a) Tufting robots and (b) tufting head

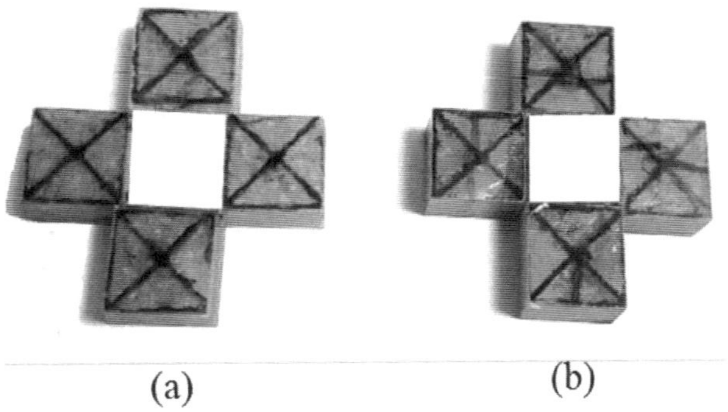

Figure 2. Manufactured tufted laminates: (a) quasi-isotropic and (b) orthotropic

2.2. *Tensile experiment*

This section explains how to set up a pulling experiment on the tufted laminates with a tensile machine equipped with a 5 KN load cell [7]. As shown in Figure 3, the samples are mounted with epoxy on two aluminium T-shape parts fixed to the top and bottom of the tensile machine. To prevent bending of the samples during the pulling test, a plastic support helps the sample to be in the centre of the aluminium parts. Four tensile tests are performed for each laminate type. Before performing the tensile tests, some tufted samples without release films are also loaded in tension to estimate the compliance of the testing set-up and the aluminium block to obtain the accurate tuft stiffness properties.

Figure 3. Tensile test setup

3. Results and discussion

The tensile experiment results are displayed in Figures 4(a) and (b) for both laminate types. Orthotropic laminates are labeled L1 while quasi-isotropic laminates are labeled L2. The damaged samples show that the tuft failed in the pre-crack layer in quasi-isotropic laminate, while a small tuft pulling out can be observed in orthotropic laminates. The CT-scan pictures of laminates tope view cross-section are shown in Figures 5(a) and (b) for two different sequences. The CT-scan observation shows that the tuft is kept tighter within quasi-isotropic laminate due to less resin around the tuft against orthotropic one with a more considerable thickness of resin around the tuft. It can be understood that the tighter the tuft is inside the laminate, the less chance the tuft has to be pulled out from a quasi-isotropic laminate than an orthotropic one.

The mean force-displacement curves are plotted in Figures 4(a) and (b) for each laminate type. The results show that the ultimate tuft tensile strength of the quasi-isotropic layup is almost 1.6 times higher than the orthotropic laminate. On the other hand, after reaching the maximum force, the orthotropic diagram decreases progressively when compared to the quasi-isotropic diagram that drops abruptly (total displacement of 1.3mm for orthotropic laminates in comparison with 0.3mm for quasi-isotropic laminate). This difference in failure mechanism yields a significantly higher fracture energy for orthotropic laminates where some tuft pull-out is visible on the fractured surface.

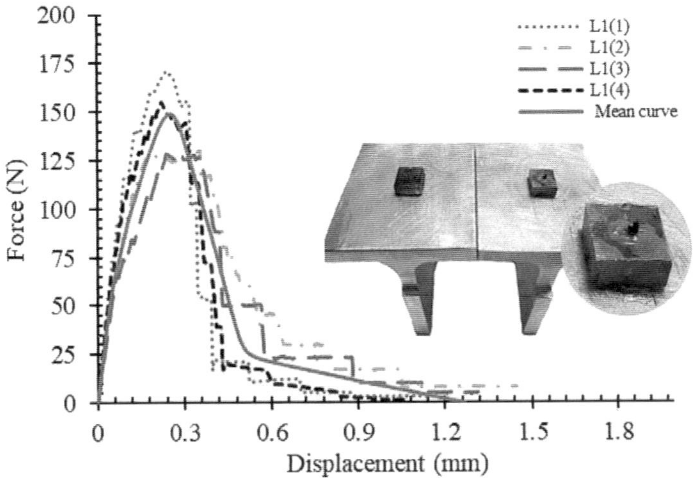

(a)

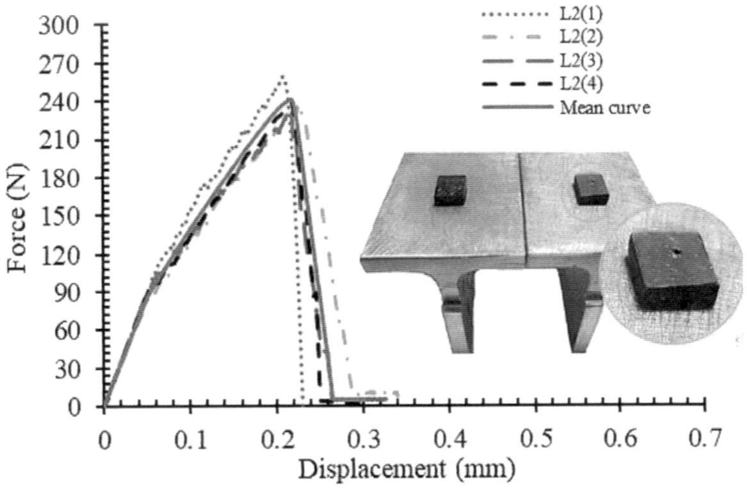

(b)

Figure 4. Force-displacement diagrams: (a) orthotropic and (b) quasi-isotropic laminate

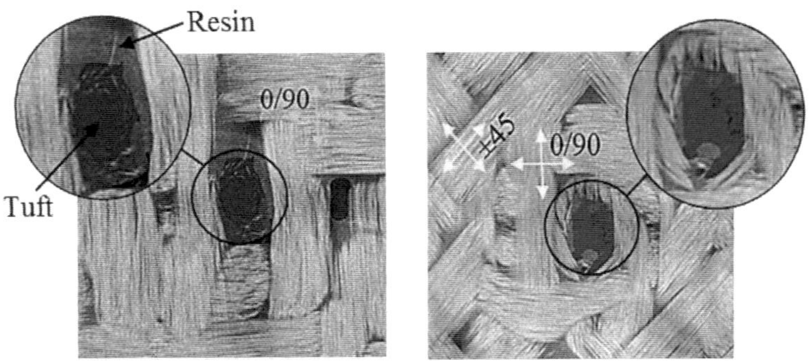

(a) (b)

Figure 5. CT-scan top view pictures of (a) orthotropic and (b) quasi-isotropic laminates

4. Conclusion

This study focuses on the tuft bridging in GFRP laminates subjected to uniaxial pulling tests. The preforms are twill 2/2 woven E-glass fabrics with two different layups. From the pulling test findings reported herein, the following conclusions are drawn:

- The amount of resin surrounding the tuft can significantly influence tuft properties when subjected to tensile loading. When a small amount of resin is present and the tuft is in close contact with the laminate, tuft failure at the mid-plane is observed. Conversely, a large amount of resin around the tuft favours fiber pull-out.

- The quasi-isotropic laminate, which have a relatively small amount of resin surrounding the tuft, have a 60% higher tensile strength than orthotropic laminates that have a larger amount of resin around tufts.

- The fracture energy of tufted orthotropic laminates in which the tuft pull-out is visible on fractured surface is 62% higher than the fracture energy of tufted quasi-iso laminates.

References

[1] Osmiani C, Mohamed G, Treiber J.W.G, Allegri, G, Partridge I.K 2016 *Exploring the influence of micro-structure on the mechanical properties and crack bridging mechanisms of fibrous tufts* (Composites Part A: Applied Science and Manufacturing vol 91) pp 409-419.

[2] Leduc C, Joncas S, Leclerc J *2014 Effects of one-side stitching parameters on damage tolerance of carbon-epoxy laminates Int. Conf.* ECCM-16 (Seville, Spain)

[3] Blacklock M, Joosten M, Pingkarawat K, Mouritz AP 2016 *Prediction of mode I delamination resistance of z-pinned laminates using the embedded finite element technique* (Composites Part A: Applied Science and Manufacturing vol 91) pp 283–91.

[4] Dell'Anno G, Treiber J.W.G, Partridge I.K 2016 *Manufacturing of composite parts reinforced through-thickness by tufting* (Robotics and Computer-Integrated Manufacturing vol 37) pp 262–272

[5] Treiber J.W.G, Cartié D.D.R, Partridge I.K 2009 *Determination of crack bridging laws in tufted composites*

[6] Pappas G, Joncas S, Michaud V, Botsis J 2018 *The influence of through-thickness reinforcement geometry and pattern on delamination of fiber-reinforced composites: Part I – Experimental results* (composite structure vol 184) pp 924–934.

[7] Pappas G, Joncas S, Michaud V, Botsis J 2017 *The influence of through-thickness reinforcement geometry and pattern on delamination of fiber-reinforced composites: Part II – Modeling* (composite structure vol 181) pp 379–390.

IOP Conf. Series: Materials Science and Engineering 1266 (2023) 012011 doi:10.1088/1757-899X/1266/1/012011

Morphological study on composite materials developed through reinforcing natural and synthetic woven fabrics from glass and hemp

K. M. Faridul Hasan *1, György Török 1, Chenar Ali Tahir 2, Péter György Horváth 2, Miklós Bak 2, and Tibor Alpár *3

[1]Simonyi Károly Faculty of Engineering, University of Sopron, Sopron, Hungary

[2]Department of Timber Architecture, University of Sopron, Sopron, Hungary

Corresponding authors E-mail: faridulwtu@outlook.com (K. M. F. H.) and alpar.tibor@uni-sopron.hu (T. A.)

Abstract. The composites made of natural and synthetic fiber reinforced materials are getting attentions continuously for different engineering applications. Previously, only synthetic fibers were considered for the reinforcement materials due to their superior mechanical properties. However, with the span of time natural fibers are also gaining popularity for their sustainable features. However, the pretreatment of fiber materials could enhance the thermomechanical performances through improving the fiber to matrix interfaces. In this regard, a comparative study is conducted between the synthetic glass and natural hemp woven fabrics to investigate their morphological (before and after the tensile loads) properties. Furthermore, the mechanical tensile properties (tensile and flexural) also shown improved mechanical performances of the products. Glass fiber reinforced epoxy composite shown 79 (1.8) MPa tensile strength, whereas the hemp fibers reinforced composites only provided 39 (1.5) MPa. In case of flexural characteristics, glass fibers also showing better strength by 196 (32.8) MPa than that of hemp 48 (3.5). Thermal stability of the products was also tested using TGA (Thermogravimetric analysis)/DTG (Derivative thermogravimetry) analysis and found that glass fiber reinforced composites have better stability than that of hemp. The results obtained from the developed composite materials clearly reflects the significant differences between the two types of woven fabrics.

1. Introduction

In current times, there is a significant interest growing on natural fiber reinforced sustainable biocomposites. However, there is also some efforts going on to compare the properties of natural and synthetic fiber reinforced composite materials. As the hemp is also a natural fiber [1], hence it is becoming a potential candidate for composite materials development. On the other hand, glass is a most popularly used materials for composite too [2, 3]. However, some efforts are also going on to improve their thermomechanical characteristics. The compatibility of the woven reinforcements could play an important role for enhancing the fiber to matrix interactions which could facilitate the improved thermo-mechanical properties from the developed materials [4-6]. It is tried in this study to develop the fiber to matrix interaction in composite system through applying the pretreatment to the fiber materials.

The interfaces are generated when the fibers with higher strength and stiffness are reinforced with a polymer matrix in composite system. There are several treatment process available to improve the fiber to matrix interfaces [7, 8]. Alkaline treatment which is also termed as mercerization treatment as well is a popularly used natural fiber treatment process [9]. The woven flax fabric was treated with alkaline solutions for improving the fiber to matrix performance characteristics. A proposed reaction mechanism is shown in equation **(1)**. The alkaline treatment reduces the diameter of fiber through increasing the fiber aspect ratio thus develops a rougher surface topography which enables improved mechanical performances of natural fiber reinforced biocomposites. The possibility of reaction is also increased through the fiber treatment which consequences better fiber to matrix bonding too [10]. The mercerization process also further influences the molecular orientation and degree of polymerization of cellulose crystallinity in hemp for the removal of cementing polymers like hemicellulose and lignin [10].

$$Hemp\ woven\ fabric - OH + Na - OH \rightarrow Hemp\ woven\ fabric - O - Na + H_2O \quad (1)$$

Conversely, glass fiber treatment by vinyltrimithoxysilane to improve the glassy surface through generating a functional film to transfer stress from polymer to fiber through increasing fiber wettability and compatibility [11]. In this regard, the woven glass fabric was treated by silanes to improve the fiber adhesion property in composite matrix system.

2. Materials and methods

The glass woven fabrics (plain weave, grid size: 4.4 mm × 4.5 mm) having 230 g/m^2 density were bought from Tolnatext company in Hungary. The hemp woven fabric was purchased from Rambutan Vietnam. The epoxy resin (CAS 25068-38-6) and hardener (epoxy katalizator, Cas-Szam 112-24-3) were collected from Novia Kft. (Romania). The mixing ratio of epoxy resin:hardener was 100:35.

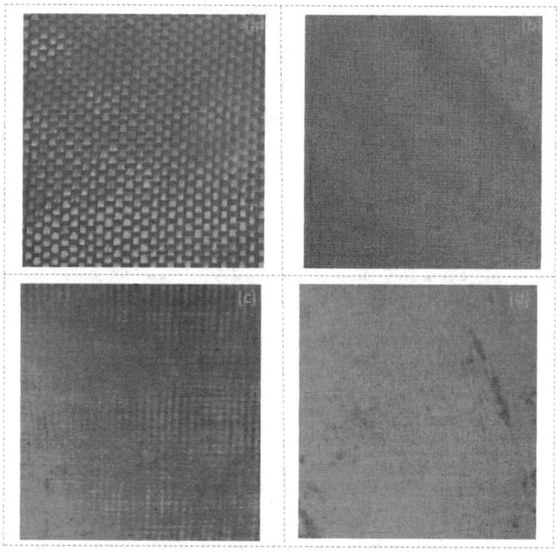

Figure 1. Photographs of developed control fibers and their composites at: (a) glass fabric, (b) hemp fabric, (b and (c) glass fabric reinforced epoxy composite, and (d) hemp fabric epoxy reinforced composite.

The glass and hemp woven fabrics were pretreated with vinyltrimithoxysilane and alkaline NaOH solution (5%), respectively. The glass pretreatment was performed at room temperature, whereas the hemp was pretreated at 100 °C for 45 min. The fabric samples were then dried at 80 °C for 7 min. Later, four layers of hemp and four layers of glass woven fabrics were prepared by 400 × 400 mm^2 dimensions. Initially, the glass woven fabrics were stacked one over another through coating with epoxy resin and hardener mixture uniformly. The hemp woven fabrics were also laminated similarly. The laminated composites were then dried at room temperature for 24 h to ensure better curations (). The SEM (Scanning electron microscopy, Hitachi S 3400 N, Japan) characterization of the developed composites were performed. Mechanical properties were performed using Instron machine (United States). TGA analysis was conducted using Themys thermal analyzer (Setaram instrumentation, France) within 25 to 750 °C under nitrogen atmosphere at 10 °C/min.

3. Results and discussion

The SEM images of the developed composites are shown in Figure 2 which represents the morphological structure of the fractured and unfractured surfaces. The photographs before applying the tensile loads do not show the appearance of respective fibers like glass/hemp as only surfaces could be seen here, which is maybe for the strong coating of the epoxy resin on the surfaces.

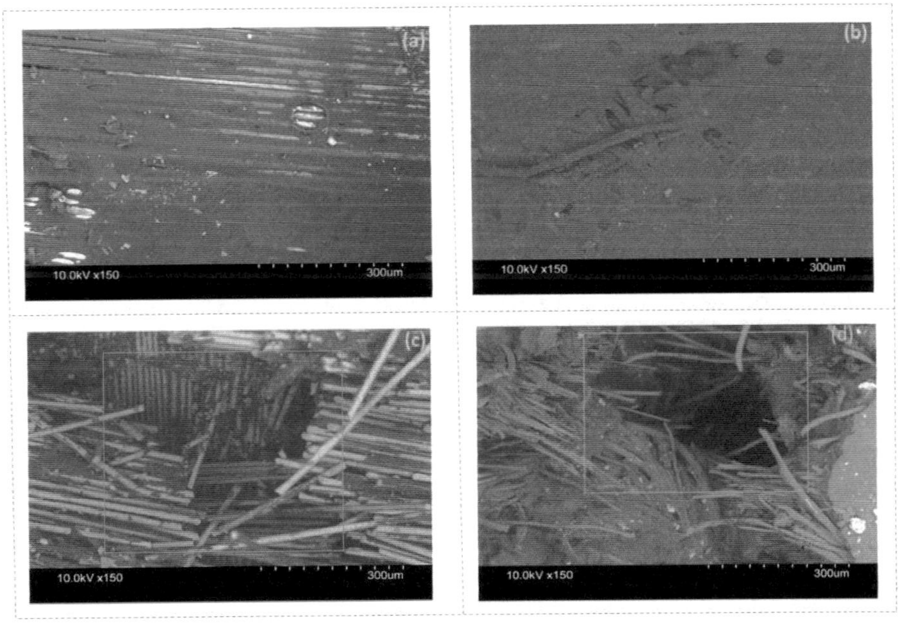

Figure 2. SEM morphologies of developed composites at different magnifications before and after fracture: (a and c) glass/epoxy and (b and d) hemp/epoxy.

However, the fractured SEM images shown the existence of fibers in the composite matrix system. The glass woven fabrics are biaxial, hence less adhesive contact is observed within the stacked layers which is prone to more delamination upon the tensile loads. Conversely, amount of resin for glass fabrics also found lower compared to flax woven fabrics as seen in Figure 2 (a and b). The phenomenon also agreed with previous study by other researcher [9, 12]. It is further found that bonding between hemp and epoxy is higher than the glass/epoxy composites. In this regard, it could be summarized that hemp/glass with epoxy composite could enhance the better adhesion property between the reinforcements and polymer to form hybrid composites with better performances.

Moreover, mechanical properties were also investigated in terms of tensile and flexural performances (Table 1). It is seen that glass fibers reinforced composites shown higher tensile and flexural properties than the hemp fibers. The tensile strength shown by the glass fiber composite is 79 (1.8) MPa, whereas hemp fiber reinforced composites shown 39 (1.5) MPa only. Similarly flexural properties shown by glass and hemp fiber reinforced composites are 196 (32.8) and 48.9 (3.5) MPa, respectively. Youngs modulus also shown the similar effects. The results found here also goes in line with some other researches for flax/glass fiber reinforced polymeric composites [2, 13]. It maybe that, the mercerization of the fibers provided an improved mechanical property due to the strong reinforcement effect between the fibers and polymers as seen in SEM images (Figure 2).

Table 1. Mechanical properties of developed composites.

	Tensile strength (MPa)	Youngs modulus (GPa)	Flexural strength (MPa)	Youngs modulus (GPa)
100% glass/epoxy laminates	79 (1.8)	7 (0.05)	196 (32.8)	53 (3.2)
100% hemp/epoxy laminates	39 (1.5)	3 (0.06)	48 (3.5)	5 (0.25)

TGA and DTG analysis were also conducted to investigate the thermal stability of the composites (Figure 3) further. It is seen that, 100% glass woven fabric reinforced epoxy composite display better thermal stability compared to 100% hemp woven fabric reinforced epoxy composites. It maybe that the glass fibers are more thermally stable compared to natural hemp fibers [14]. DTG analysis demonstrates that about the presence of glassy materials (Figure 3), cellulose, hemicellulose, and lignin in the products. The composite products initially losses some weight (3 to 5%) until 100 °C maybe due to the presence of moisture [15, 16]. However, glass fibers did not show this trend like as the hemp fibers reinforced composites. Therefore, hemp fibers losses more weight especially after 300 °C.

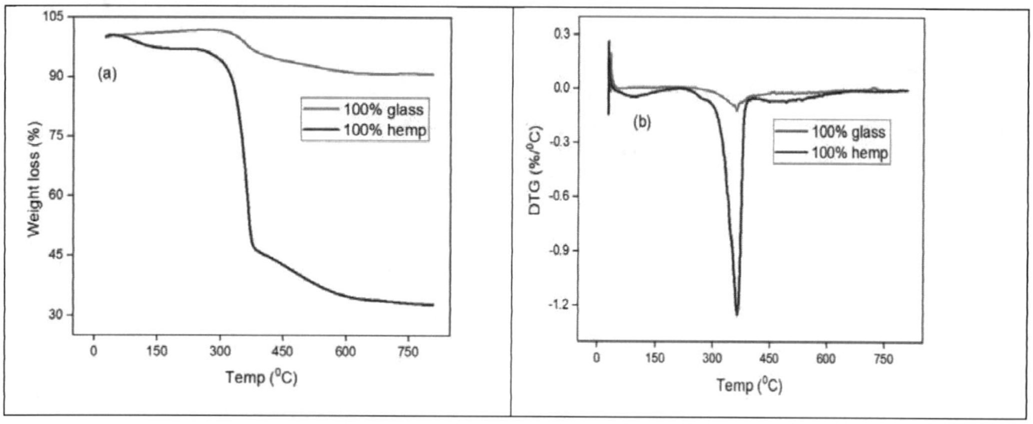

Figure 3. Thermal properties analysis: (a) TGA and (b) DTG.

4. Conclusion

The glass and hemp woven fabric reinforced composites showed significant differences between the morphological properties. The pretreatment has significantly improved the fiber to matrix adhesion properties. Although the unfractured composites did not show the presences of glass and hemp fibers on the laminated composites but the fractured surfaces showed the explicit presence of them. Tensile and flexural properties also found impressive, whereas glass fiber reinforced composites shown higher performances than that of flax. Additionally, glass fiber reinforced composites also displayed higher thermal stability than that of hemp. However, further researches are needed to conduct for developing some hybrid composites through reinforcing both the hemp and glass woven fabrics together with the polymeric matrix for investigating their thermomechanical performances. Besides, the developed materials could be used safely against crack propagation through minimizing the environmental pollutions. The inferred results also further clarified the class of materials in terms of mechanical and fracture characteristics.

Acknowledgment

This work was kindly supported by the “Stipendium Hungaricum” grant under the Simonyi Károly Faculty of Engineering, Wood Sciences, and Applied Arts, University of Sopron, Hungary. This article was produced within the framework of “TKP2021-NKTA-43” project with the support provided by the Ministry of Innovation and Technology of Hungary from the National Research, Development and Innovation Fund, financed under the TKP2021-NKTA funding scheme.

References

[1] Hasan K M F, Horváth P G and Alpár T 2020 Potential Natural Fiber Polymeric Nanobiocomposites: A Review *Polymers* **12** 1072
[2] Morye S and Wool R 2005 Mechanical properties of glass/flax hybrid composites based on a novel modified soybean oil matrix material *Polym. Compos* **26** 407-16
[3] Sherif G, Chukov D, Tcherdyntsev V and Torokhov V 2019 Effect of formation route on the mechanical properties of the polyethersulfone composites reinforced with glass fibers *Polymers* **11** 1364
[4] Hasan K M F, Péter G H, Gábor M and Tibor A 2021 Thermo-mechanical characteristics of flax woven fabric reinforced PLA and PP biocomposites *Green Mater* 1-9
[5] Fiore V and Calabrese L 2019 Effect of stacking sequence and sodium bicarbonate treatment on quasi-static and dynamic mechanical properties of flax/jute epoxy-based composites *Materials* **12** 1363
[6] Hasan K M F, Horváth P G and Alpár T 2020 Effects of alkaline treatments on coconut fiber reinforced biocomposites In: *9th Interdisciplinary Doctoral Conference,* ed H C Dr. Csiszár Beáta, Kajos Luca Fanni, Kovács Olivér Béla, Mez˝o Emerencia, Szabó Rebeka, Szabó-Guth Kitti (Pecs, Hungary: Doctoral Student Association of the University of Pécs) p 248
[7] Koohestani B, Darban A, Mokhtari P, Yilmaz E and Darezereshki E 2019 Comparison of different natural fiber treatments: A literature review *Int. J. Environ. Sci. Technol* **16** 629-42
[8] Mahmud S, Hasan K F, Jahid M A, Mohiuddin K, Zhang R and Zhu J 2021 Comprehensive review on plant fiber-reinforced polymeric biocomposites *J. Mater. Sci* 1-34
[9] Hasan K M F, Péter György H and Tibor A 2020 Thermo-mechanical behavior of MDI bonded flax/glass woven fabric reinforced laminated composites *ACS Omega*
[10] Kalia S, Kaith B and Kaur I 2009 Pretreatments of natural fibers and their application as reinforcing material in polymer composites—a review *Polym. Eng. Sci* **49** 1253-72
[11] Cech V, Prikryl R, Balkova R, Vanek J and Grycova A 2003 The influence of surface modifications of glass on glass fiber/polyester interphase properties *J. Adhes. Sci. Technol* **17** 1299-320
[12] Hasan K F, Horváth P G, Zsolt K, Kóczán Z, Bak M, Horváth A and Alpár T 2021 Hemp/glass woven fabric reinforced laminated nanocomposites via in-situ synthesized silver nanoparticles from Tilia cordata leaf extract *Compos Interfaces* 1-19
[13] Ramesh M and Nijanthan S 2016 Mechanical property analysis of kenaf–glass fibre reinforced polymer composites using finite element analysis *Bull. Mater. Sci* **39** 147-57
[14] Hasan K F, Horváth P t G r and Alpár T 2021 Thermomechanical Behavior of Methylene Diphenyl Diisocyanate-Bonded Flax/Glass Woven Fabric Reinforced Laminated Composites *ACS Omega* **6** 6124–33
[15] Gargol M, Klepka T, Klapiszewski Ł and Podkościelna B 2021 Synthesis and thermo-mechanical study of epoxy resin-based composites with waste fibers of hemp as an eco-friendly filler *Polymers* **13** 503

[16] Rodríguez E, Petrucci R, Puglia D, Kenny J M and Vazquez A 2005 Characterization of composites based on natural and glass fibers obtained by vacuum infusion *J. Compos. Mater* **39** 265-82

IOP Conf. Series: Materials Science and Engineering 1266 (2023) 012012 doi:10.1088/1757-899X/1266/1/012012

Manufacturing of near-net-shape composite parts using Tailored Fiber Placement preforms

Marie-Claude Bélanger[1], Patricia Forcier[2], Alain Bujold[3], Simon Pesant[3], Serge Pagé[4]

[1]Centre de développement des composites du Québec, Cégep de St-Jerome, Canada, (mcbelang@cstj.qc.ca)
[2]CTT Group, Cégep de St-Hyacinthe, Canada (pforcier@gcttg.com)
[3]Mawashi Science and Technology, Canada (abujold@mawashi.net; spesant@mawashi.net)
[4]Mat-Comp, Canada (spage@mat-comp.ca)

Abstract. This paper demonstrates the feasibility of manufacturing a light and low cost near-net-shape vertebra from Tailored Fiber Placement (TFP) preforms. The TFP preforms were produced flat and sewn together at the ends by a classic sewing process. Low pressure Vacuum Assisted Resin Transfer Molding (VARTM) was used for the consolidation and impregnation of the preforms. The molding strategy was oriented towards the use of silicone conformers to ensure a compaction that well fits the preform despite its imperfections. The resulting part required a light finishing. The mass was reduced by 40% and the manufacturing cost by 76% compared to the original part.

1. Introduction:

Developed in the 1990s, Tailored Fiber Placement (TFP) is based on basic embroidery technologies and can manufacture near-net-shape and net-shape textile preforms for composite parts with fiber layouts in any direction. The principle is shown in Fig. 1. A continuous fiber roving is fixed on the support material by a zigzag lock stitch with an upper and a lower thread. TFP provides great design flexibility because it allows generating a variety of shapes, configurations and thicknesses, while imposing the required direction of reinforcements. TFP is automated, consistent and repeatable. The combination of various fiber, base materials and sewing thread is possible, depending on the specifications and standards to be met and depending on the end use of the product.

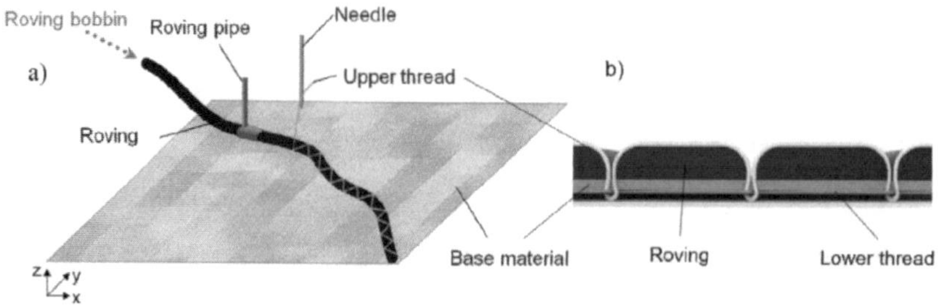

Figure 1. Principle of TFP process. a) Roving being place on the base material and fixed using a zigzag stitch (Wikipedia). b) Cross section showing a modified lock stitch stitching method [6].

It has been shown in a previous study [1] that TFP using Kevlar® thread provides a 10 % higher impact resistance versus polyester thread and has a better impact resistance than a stitchless baseline configuration. Also, using Kevlar® thread in TFP increases the resistance to delamination by 38 %. However, the in-plane shear strength is reduced by 24 % due to the damage incurred as a result of the stitching process [2, 3, 4,7].

Thus, composites made of TFP can be considered as candidate materials for structures that sustain significant impact damage (e.g. due to dropped tools and ballistic loads) and high peel stresses (e.g. in joints), and structures where cut-outs (e.g. edges and holes) [5] are difficult to avoid.

The objective of the present paper is to demonstrate the feasibility of manufacturing a light and low cost near-net-shape vertebra from a TFP preform. The vertebra is part of the Mawashi's UPRISE™ exoskeleton: a complex structure made of high strength titanium that allows carrying up to 120 lbs with ease while maintaining a great range of motion. The agility of the system is that it is based on the movements of the human body, and it protects it from common injuries due to load carrying.

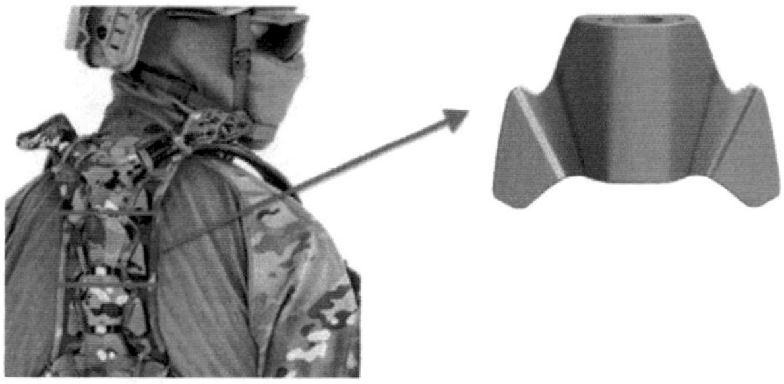

Figure 2. Vertebra from UPRISE^{MC} exoskeleton (Mawashi.net), currently made of 3D-printed titanium alloy

The use of composite materials instead of printed titanium would make the part lighter and would reduce its manufacturing cost. However, manufacturing such a part in composite material using common processes is a challenge. The placement of reinforcements is complex due to the small size of the part and any secondary operations (cutting, assembly, finishing) will add additional costs. The manufacture of embroidered preforms by TFP can overcome these obstacles.

2. Material and Methods

TFP preforms were manufactured using an embroidery machine with a head designed for tailored fiber placement. The stacking configuration was chosen using the results of a previous study [1] that investigated the effects of various thread materials and configurations on the in-plane shear strength as well as on the damage resistance and compression after impact strength of TFP laminates.

IOP Conf. Series: Materials Science and Engineering 1266 (2023) 012012 doi:10.1088/1757-899X/1266/1/012012

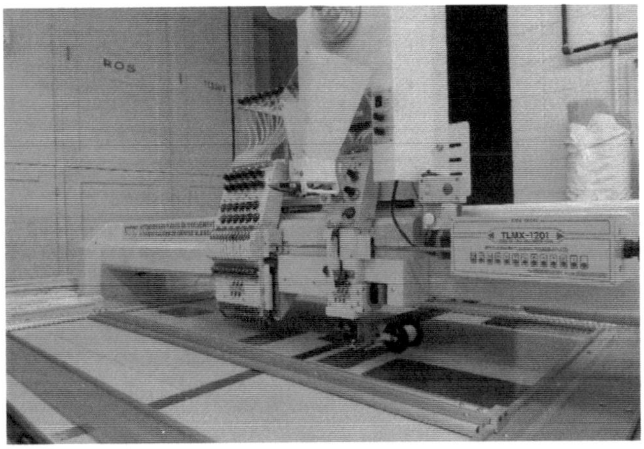

Figure 3. Tajima enbroidery machine at CTT Group, with TFP device.

The design of preforms by TFP raised several challenges. First, the choice of the base substrate had to be in the same material as the threads deposited: being in carbon, a veil of the same composition was selected. Any other substrate commonly used in the conventional embroidery world, such as polyester, would have been considered a contaminant. In addition, the residual quantity of this substrate had to be at a minimum, hence the choice of a very light veil. Given the large number of zig-zag points generated by the production of the preforms, the substrate was doubled in order to overcome the problem of tearing generated on a single layer.

Wanting to reduce as much as possible the number of points necessary to make the preforms, a wick of HTS40-12K carbon was selected. Thus, a single segment of this roving had higher coverage than a smaller roving. However, the smallest details in the shape and design of the vertebra were more difficult to reproduce. In consequence, the contour of the preform tended to lose its uniformity. A smaller roving type would certainly help alleviate this issue but would substantially increase the number of zig-zag stitches.

For each vertebra produced, two TFP preforms were required. These preforms were made flat on the embroidery machine, in accordance with the thicknesses and reinforcements required. To compensate for potential empty zones that would create resin pockets between the two preforms, some areas were reinforced with fibers.

Following the TFP process, a cleaning and inspection of the preforms were carried out. The preforms were placed on a template to perform the first 3D shaping step. They were then sewn together at the ends using a classic sewing process to maintain and secure their shape (figure 4). To avoid contamination, the sewing thread used was the same as the type used during the TFP process.

Figure 4. Final shape of the perform

IOP Conf. Series: Materials Science and Engineering 1266 (2023) 012012 doi:10.1088/1757-899X/1266/1/012012

Inserts were added to the preform before impregnation. One of these inserts is made of porous 3D-printed polyetherimide and acts as a core to lighten the part. The other insert is made of printed titanium and is a link to the other vertebrae of the exoskeleton.

Low pressure Vacuum Assisted Resin Transfer Molding (VARTM) was used for the consolidation and impregnation of the preforms. Silicone conformers were used during the molding process to ensure a compaction level that conforms the preform well despite its imperfections. In addition, the resin was directed through integrated undercut channels. The silicone conformers were manufactured by casting in molds printed to the desired final geometry. The conformers were placed in printed lids, whose main function was to ensure air tightness. Prior to molding, the prefabricated inserts were placed and positioned by the silicone conformers. Figure 5 shows the preform (black) placed in the male conformer (green). This one is placed in its cover (grey).

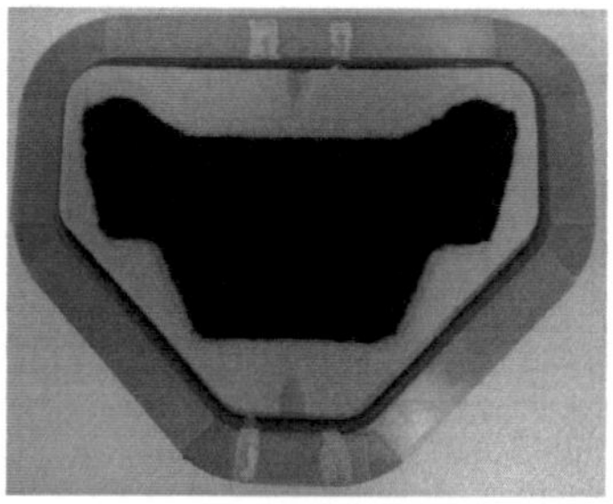

Figure 5. TFP preform placed in the male conformer

The resin system was composed of Adtech 820 low viscosity epoxy with 824 hardener. Injection pressure was gradually raised from atmospheric pressure to 5 psig. Mould temperature was kept at 25 °C throughout the injection process. Twenty-four hours after hardening, the vertebra was demolded and post cured. The resulting part required a light finishing to remove resin channels.

Figure 6. Final part

3. Results and discussion

The assembly of the flat preforms was a challenge. They were difficult to hold in place even with a jig. Since the assembly is done using a classical sewing process, it has to be strong and precise enough for a final preform of 5 mm thickness. Therefore, a lockstitch zig-zag seam with a needle usually used for jeans (size 16) is recommended on the sewing machine, even if this type of stitch increases the number of stitches, and even if a needle of this size increases the risk of damaging the fibers.

The stitch pattern caused a slight wave in the rovings, which introduced misalignment of the in-plane fibers, a thread fiber crimp and the formation of resin pockets. Also, the contour of the preform tended to be resin rich and difficult to control. Special care was needed when placing the stacked preforms into the mold, to ensure proper closing and sealing.

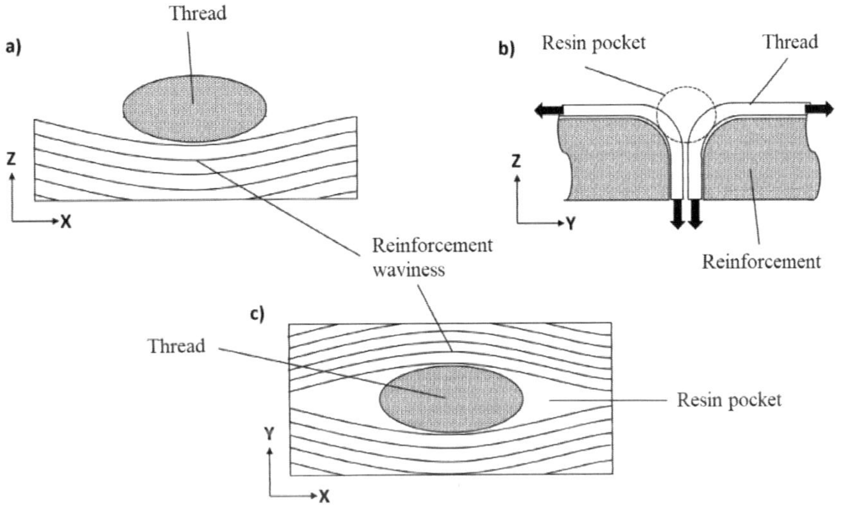

Figure 7. Typical defects caused by the insertion of threads: a) waviness of out-of-plane reinforcement, b) resin-rich area at the insertion point and c) waviness and resin-rich area in the plane [8].

The molding process (VARTM) made high reinforcement volume and controlled surfaces possible. However, it is sensitive to variations in the perform assembly. Design improvements were necessary to ensure the closing and the sealing of the mold, and to increase the compaction of the preforms, in order to increase the fiber ratio.

After demolding, the final part was near-net-shape (negligible finishing). The carbon fiber volume ratio was estimated between 45 and 49%. The mass was reduced by 40% and the manufacturing cost by 76% compared to the original part. The proposed solution and manufacturing strategy have been validated.

4. Conclusion

Near-net-shape composite parts were manufactured using TFP preforms. The preforms were assembled and impregnated using low pressure VARTM, resulting in a high reinforcement carbon fiber volume ratio. Some defects were inherent to the TFP technology. For example, waviness of out-of-plane reinforcement and resin-rich area at the insertion point of the needle. In general, the final part showed small imperfections.

Nevertheless, the automated embroidery and the proposed molding strategy have shown to significantly reduce the mass and production cost of the vertebra compared to the printed titanium model

Acknowledgment: The research presented herein has been made possible by funding from the National Science and Engineering Research Council of Canada (NSERC) and by the contribution of industrial partners: Mawashi Science and Technology and Mat-Comp.

References:

[1] M.-C. Bélanger, P. Forcier, Y. Gendreau, A. Bujold, S. Pesant, S. Pagé and L. Laberge Lebel, "Effect of tailored fiber placement threads on carbon/epoxy composite impact resistance" *CAM-X 2019 – The Composites and Advanced Materials Expo (Conference Proceedings)*, Anaheim, USA, September 23-26 2019.

[2] P. Mattheij, K. Gliesche, and D. Feltin, "3D reinforced stitched carbon/epoxy laminates made by tailored fibre placement," *Composites Part A: Applied Science and Manufacturing,* vol. 31, no. 6, pp. 571-581, 2000.

[3] A. Mouritz, K. Leong, and I. Herszberg, "A review of the effect of stitching on the in-plane mechanical properties of fibre-reinforced polymer composites," *Composites Part A: Applied Science and Manufacturing,* vol. 28, no. 12, pp. 979-991, 1997.

[4] H. Heß and N. Himmel, "Structurally stitched NCF CFRP laminates. Part 1: Experimental characterization of in-plane and out-of-plane properties," *Composites Science and Technology,* vol. 71, no. 5, pp. 549-568, 2011.

[5] E. G. Koricho, A. Khomenko, T. Fristedt, and M. Haq, "Innovative tailored fiber placement technique for enhanced damage resistance in notched composite laminate," *Composite Structures,* vol. 120, pp. 378-385, 2015.

[6] Y. Gendreau, M.-C. Bélanger, J. Lévesque and L. Laberge Lebel. "Effect of Tailored Fiber Placement Parameters on Carbon Fiber/Epoxy Composite," *Canada-Japan Workshop on Composites (Conference Proceedings)*, Takayama, Japan, July 4-7 2018.

[7] P. Mattheij, K. Gliesche, and D. Feltin, "Tailored Fiber Placement-Mechanical Properties and Applications," *Journal of Reinforced Plastics and Composites,* vol. 17, no. 9, pp. 774-786, June 1st 1998.

[8] Y. Gendreau (2019). "Modélisation et simulation de composite à renfort fabriqué par broderie," *Master's thesis*, Polytechnique Montréal.

Investigation on bio-sourced textile reinforcement for composite material based on sisal Moroccan yarns

Z Samouh[1,2,3], A Abed[1,2,3], C Cochrane[1], A R Labanieh[1], F Boussu[*1], D Soulat[1], R El-Mozznine[2], O Cherkaoui[3]

[1] GEMTEX Laboratory, ENSAIT, Univ. Lille North of France, F-59100 Roubaix, France

[2] LPMC Laboratory, Faculty of Science El Jadida, Chouaib Doukkali University, El Jadida, Morocco

[3] REMTEX Laboratory, ESITH (Higher School of Textile and Clothing Industries), Casablanca, Morocco

francois.boussu@ensait.fr

Abstract. The main objective of this paper aims at investigating the potential use of sisal yarn into composite material despite the inherent variability of properties of natural resources. A multi-scale approach of the behavior of sisal fiber woven reinforcements is conducted to understand and evaluate the different properties of woven reinforcements. At the yarn scale, a piezo-resistive sensor yarn was developed to assess deformations and stress concentrations in-situ in order to understand the material behavior during the weaving of woven reinforcements fibrous for bio-sourced composite materials. At the fabric scale, 2D woven reinforcements are developed based on a conventional weaving process. The production and characterization of composite sheets based on 2D woven reinforcements show the potential of sisal fiber woven reinforcements compared to natural fiber woven reinforcements from literature.

1 Introduction

The next generation of composite materials will contain more and more bio-sourced materials, both for the matrix and for the fibrous reinforcement. Since the last past ten years, several research works have investigated the different types of natural resources to be used [1][2][3] for the textile reinforcement. Among these materials, sisal yarns produced in Morocco appears as a sustainable resource. Several mechanical and physical characterizations have been done at the filament and the yarn scales and it seems to be sufficiently strength to be used during the weaving process. By the same, it could be interesting to check their ability during the weaving of a 3D fabric.

Additionally, insertion of sensing devices inside the fibrous reinforcement, that could behave similarly as the structural fibrous reinforcement, could also help to monitor the effect of the resin impregnation during manufacturing process onto the residual mechanical behavior of the fibrous structure [4]. By the same, these impregnated sensing devices entirely embedded into the fibrous reinforcement could be helpful to detect some damage initiation of the composite material and then contribute to its health monitoring during its lifetime duty.

Thus, the two main objectives of this paper are: firstly, to evaluate the residual mechanical performance of 2D or 3D woven structures based on heterogeneous sisal yarns and secondly, to design,

IOP Conf. Series: Materials Science and Engineering 1266 (2023) 012013 doi:10.1088/1757-899X/1266/1/012013

produce and insert innovative piezo-resistive sensor yarns made with the sisal yarn, as the core, to replicate the same mechanical behavior as the fabric's structural yarns.

2 Material and Methods

2.1 Design and manufacturing of sisal sensor yarn

The sisal yarn was cleaned through successive sonication Soxhlet glassworks cleaning with petroleum ether as a solvent over 3 h. This cleaning process was very useful to remove all the impurities trapped inside the natural sisal yarn, which can strongly influence the interfacial adhesion properties.
The cleaned sisal yarn was pre-dipped into a PVA hydrogel for 30 seconds and then was dried at 90 °C for 15 minutes. Then, the resulting PVA-coated sisal core yarn was ligated by two copper connector wires with an inter-space value l0=30 mm. This sample was coated with four layers of piezo-resistive PEDOT:PSS polymer where each layer has been dried at 110 °C for 5 minutes. The final sensor yarn, based on sisal core and coated conductive polymer, is represented in Figure 1.

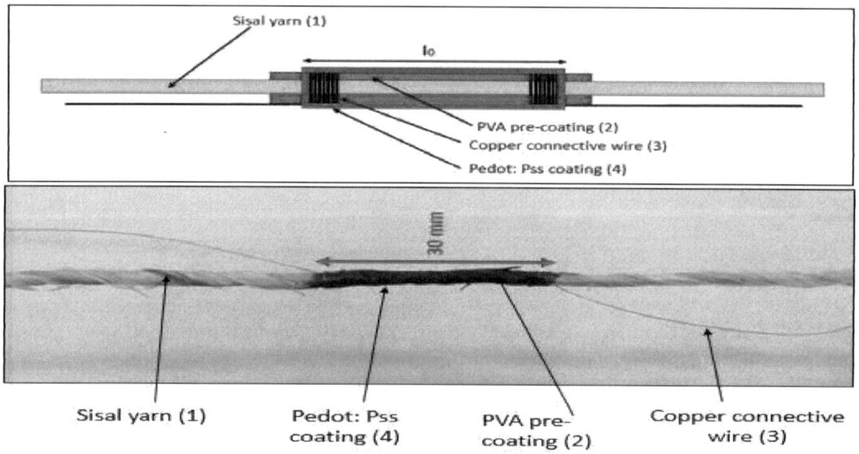

Figure 1. (a) General scheme of the sensor yarn and (b) the resulted manufactured sensor yarn.

2.2 Design and Manufacturing of 2D woven sisal fabrics

Three 2D fabrics were designed using the geometric modeling software. The fabrics were made from sisal yarn (3300 Tex) and produced on the same prototype weaving loom of the GEMTEX laboratory. After weaving, fabrics were stored at the specific room conditions (20°C, 62% HR) for 24 hours. The thickness and areal fabric density were measured according to ISO 4603 and ISO 12127 standards, respectively. The tensile properties of the fabrics have been analyzed according to the ISO 13934-1 standard. The composites based on the manufactured 2D fabrics (Table 1) are produced by vacuum infusion molding with SR8200 epoxy resin. The physical and mechanical properties of the produced composites have been analyzed according to the dedicated standards.

Table 1. Description and 3D geometrical views of the 2D woven fabrics

Fabrics	Weave diagram definition	3D Graphical Representation of woven pattern
Fabric 1	Basket 4-4	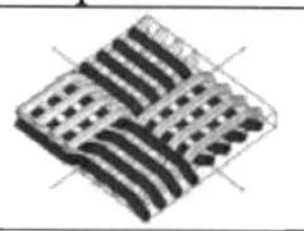

Fabric 2	4/4 Twill	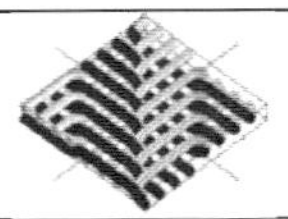
Fabric 3	8-harness satin	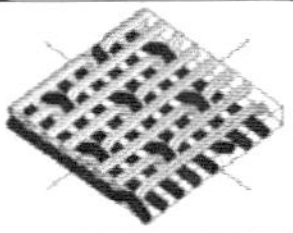

2.3 Design and Manufacturing of 2D woven sisal fabrics as fibrous reinforcement for composite material with imbedded sensor yarns

Based on the same procedure of production described previously, 60 sisal sensor yarns have been achieved. Initial resistance values of all these sisal sensor yarns have been measured. Subsequently, a calibration step was carried out by a fatigue test of 100 cycles in tension at 3% maximum elongation and at a speed of 200 mm.min-1 on a tensile test machine [5]. The resulted resistance values of all these sisal sensor yarns have again been measured, as detailed in Table 2.

Table 2. Comparison of the average resistance values of the sisal sensor yarns before and after the fatigue test.

	Initial resistance values before the fatigue test (kΩ)	Resistance values after the fatigue test (kΩ)
Average resistance	0,711	0,412
Standard deviation	0,264	0,256

Based on these sisal sensor yarns with stabilized resistance value, those with the similar resistance values have been inserted in a fabric (see description in Table 3) close to each other and both in warp and weft directions (Figure 2).

Table 3. Description of 2D woven fabric parameters (2-2 twill) with sisal yarns and imbedded sisal sensor yarns.

Product Parameters	Unit	Value
Warp yarn linear density	Tex	3300
Weft yarn linear density	Tex	3300
Average fabric thickness	mm	4,41 ± 0,23
End density	ends/cm	2 ± 0,25
Pick density	picks/cm	1,5 ± 0,25
Areal weight	kg/m2	1,18 ± 0,06

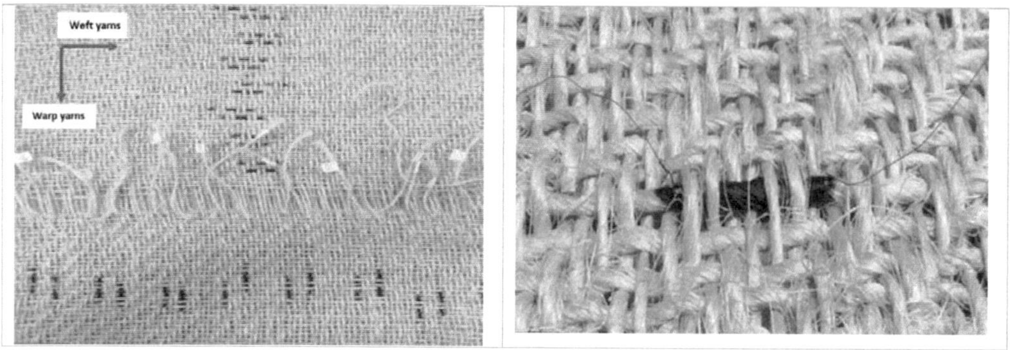

Figure 2. (left) sisal sensor yarns inserted in warp and weft directions – (right) sisal sensor yarn inserted in weft direction with metallic wires at his edges for connection

The sisal fabric with embedded sisal sensor yarns has been impregnated using vacuum infusion molding with epoxy resin to obtain the following characteristics mentioned in Table 4. For each composite coupon, two sisal sensor yarns with similar average resistance have been inserted in the weft direction to check their responses under tensile loading.

Table 4. Detailed parameters of composite material made with sisal yarns.

Parameters	Value
Average thickness (mm)	5,30 ± 0,52
Areal weight (kg/m^2)	3,694 ± 0,19
Resin volume fraction (%)	68,15 ± 1,03
Fiber volume fraction (%)	29,21 ± 2,51
Composite coupons for testing with two sisal sensor yarns inserted in the weft direction 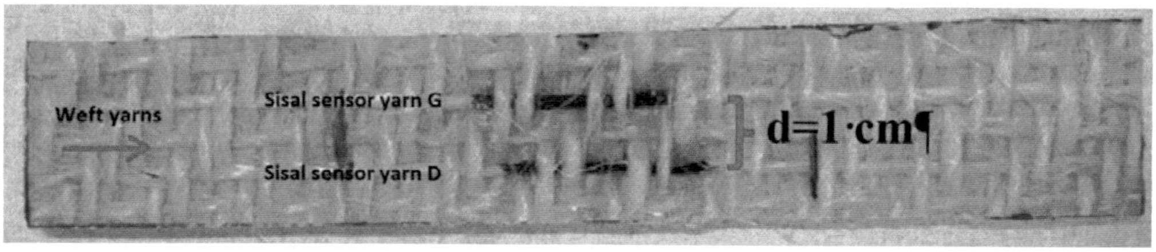	

3 Results and discussion

3.1 Characterization of the sisal sensor yarn

The piezo-resistive properties of the strain sensor yarns were investigated. Mechanical stretchings were applied on strain sensor yarn as its electrical resistance was increased and measured simultaneously. As defined in [6], the sensitive of strain sensors is determined by the gauge factor (K) through the linear relationship between the relative resistance variations ($\Delta R/R0$) with elongation as described in (Equation 1).

$$\frac{\Delta R}{R0} = K * \varepsilon \qquad \text{Equation 1}$$

Thus, the gauge factor K of a strain sensor yarn with different PEDOT:PSS loadings is calculated. It was found that the gauge factor is 1.01±0.35 and 4.75±0.59 for a strain range comprises between 0.0-3.5% and 3.5-7%, respectively for strain sensor yarn. The gauge factor is estimated to be 2.19±0.48 and 4.75±0.59 for the sensor yarns with 2 coating and 4 coating of PEDOT :PSS layers, respectively, under the strain values of 3.5-7%. Then, the strain sensor yarn with an increase amount of PEDOT:PSS layers shows higher sensitivity.

3.2 Mechanical behavior of 2D fabrics by the uniaxial tensile test

The maximum force and the deformation of 2D fabrics are shown in Table 5. The highest values of maximum force in warp and weft directions are attributed to the fabric with satin 8 weave diagram. This value can be explained, in weft direction, by the type of weave diagram, which allows having a denser structure. The elongation at break of 2D fabrics depends on the warp and weft yarns shrinkage values. After the 2D fabric composite materials, the characterization of their mechanical behavior reveals a higher value of the maximum force for the satin 8 weave diagram, in the weft direction, compared to the other samples.

Table 5. Tensile properties of fabrics and composites based on sisal yarns

Samples		Maximum Force (N) in warp Direction	Elongation at Break (%) in warp Direction	Maximum Force (N) in weft Direction	Elongation at Break (%) in weft Direction
Fabrics	Fabric 1 (Basket 4-4)	2756.65±560	4.65±0.98	4462.68±684	8.65±1.66
	Fabric 2 (4/4 Twill)	2301.54±765	5.01±1.65	3100.65±752	6.35±0.95
	Fabric 3 (8-harness satin)	3460.35±465	7.65±1.14	5425.68±826	9.36±1.97
Composites	Comp 1 (Basket 4-4)	5571.65±736	3.86±0.54	6391.58±868	3.6±0.26
	Comp 2 (4/4 Twill)	6061.36±865	3.46±0.36	6454.37±768	3.48±0.51
	Comp 3 (8-harness satin)	5867.48±456	2.4±0.52	6778.36±649	2.78±0.87

3.3 Mechanical behavior of composite coupons made with 2D sisal fabrics with imbedded sisal sensor yarns under uniaxial tensile test

Composite specimens embedded with sensor wires were tested on the MTS Criterion bench (Figure 3 left). Tensile strength tests were performed, according to ISO 527-4 in the weft direction at 10 mm.min-1, on composite specimens of nominal dimensions of 25 × 5 cm2 and tested with an edge length of 15 cm. The data acquisition system (NI USB 6003) was used to measure the strength change and the composite coupons were subjected to the tensile test to failure.

It can be observed that the two curves of the relative resistance variation of the two sisal sensor yarns seem to be quite similar, which reflects a relatively synchronous measurement by these two sisal sensor yarns. The local deformation measurements of the impregnated fabric structure therefore appear to follow the same tensile stress, which suggests a homogeneous distribution of the load applied to all the threads of the impregnated 2D sisal fabric (Figure 3 right). However, the breakage of the two sensor yarns are not strictly similar due to their initial heterogeneity. This can be highlighted by the detection of the matrix cracking at 4% of elongation, detected by the two sensor yarns, but without the same intensity (green ellipse in Figure 3 right).

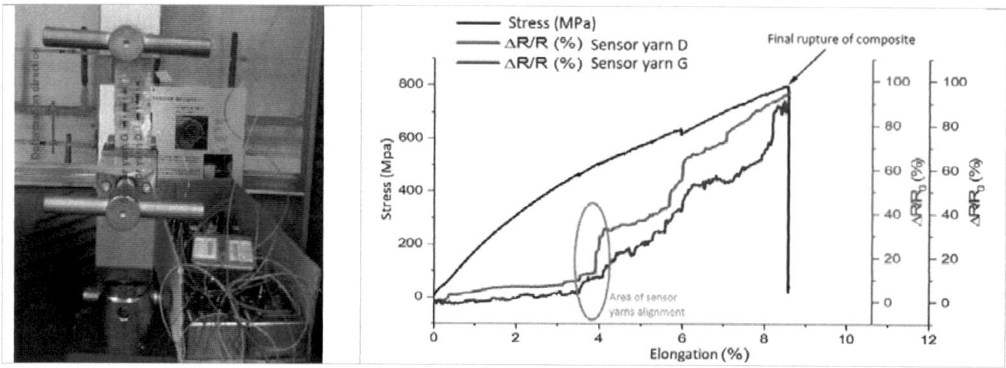

Figure 3. (left) tensile test on composite coupon of 2D sisal fabric with embedded sisal sensor yarns – (right) comparison of two sisal sensor yarns behavior

4 Conclusion

The current study reveals the potential of use of the Moroccan sisal yarn from the preparation of the reinforcement to the elaboration of the composite. These results show that the type of weave diagram can influence the mechanical performance of both the fabric and the produced composite material. Sensor yarns, based on sisal as core yarns, were developed with high sensitivity for the applied strain value comprised between 3.5 to 7%. These strain sensor yarns are produced by using a simple method, scalable and low-cost coating technique. Then, these sensors are inserted both in warp or weft directions, into the previous 2D fibrous reinforcements, to be used as local and in-situ sensing device of composite material.

5 Acknowledgment

The authors would like to express appreciation for the support of the sponsors which are the Ministry of Europe and Foreign Affairs, the Ministry of Higher Education, Research and Innovation and the French Institute of Rabat (PHC TOUBKAL 2019/83 (French-Morocco bilateral program) Grant Number: 41511ZF.

References

[1] Lansiaux H, Soulat D, Boussu F, Labanieh AR. Development and Multiscale Characterization of 3D Warp Interlock Flax Fabrics with Different Woven Architectures for Composite Applications. Fibers 2020;8. https://doi.org/10.3390/fib8020015.

[2] Corbin AC, Kececi A, Boussu F, Ferreira M, Soulat D. Engineering Design and Mechanical Property Characterisation of 3D Warp Interlock Woven Fabrics. Applied Composite Materials 2018;25:811–22. https://doi.org/10.1007/s10443-018-9715-z.

[3] Zimniewska M, Myalski J, Koziol M, Mankowski J, Bogacz E. Natural Fiber Textile Structures Suitable for Composite Materials. Journal of Natural Fibers 2012;9:229–39. https://doi.org/10.1080/15440478.2012.737176.

[4] Trifigny N, Kelly FM, Cochrane C, Boussu F, Soulat D, Koncar V. In-situ measurements of strain and stress of E-glass warp yarn during the weaving of 3D warp interlock fabric using innovative sensors. TexComp-11 Conference, Leuven, Belgique: 2013, p. 3–11.

[5] Abed A, Samouh Z, Cochrane C, Boussu F, Cherkaoui O, el Moznine R, et al. Piezo-Resistive Properties of Bio-Based Sensor Yarn Made with Sisal Fibre. Sensors 2021;21. https://doi.org/10.3390/s21124083.

[6] Huang Y, Zhao Y, Wang Y, Guo X, Zhang Y, Liu P, et al. Highly stretchable strain sensor based on polyurethane substrate using hydrogen bond-assisted laminated structure for monitoring of tiny human motions. Smart Mater Struct 2018;27:35013. https://doi.org/10.1088/1361-665x/aaaba0.

Towards commercialization of graphene-based end-of-life sensors for fire-protective fabrics

D Yehia[1], L Lawson[2], D King[3], H-J Chung[4], J Batcheller[5] and P I Dolez[6]

[1,2,3] Davey Textiles Solutions, Inc., Edmonton, AB, 10505 169 St., T5P 4Y7, Canada
[4,5,6] Department of Human Ecology & Department of Chemical and Materials Science, University of Alberta, Edmonton, AB, 114 St - 89 Ave, T6G 2E1, Canada

dyehia@daveytextiles.com, llawson@daveytextiles.com, dking@daveytextiles.com, chung3@ualberta.ca, batchell@ualberta.ca, pdolez@ualberta.ca

Abstract. Fire-resistant (FR) fabrics used in protective clothing experience a reduction in performance as a result of exposure to various ageing conditions, for instance: heat, ultraviolet (UV) light, moisture, abrasion and laundering procedures. However, there are few visible clues to indicate if the deterioration of the protective clothing has reached a dangerous level. To address this issue, graphene-based end-of-life sensors (heat, UV light, and moisture) are being developed at the University of Alberta in collaboration with five industry partners, including Davey Textile Solutions, Inc (DTS). DTS has the production capacity to manufacture the graphene-based end-of-life sensors, including weaving, finishing, conductive track application, fusing, and product assembly. The lifetime of fire-protective clothing is an important parameter to monitor, and the graphene-based end-of-life sensors are a straightforward, non-destructive, and effective tool for this purpose. The plan is to fabricate, integrate and commercialize the sensors. DTS, alongside academic researchers from the University of Alberta, are in the process of scaling up the manufacturing and testing of the sensors. The health and safety of firefighters will be improved by bringing graphene-based end-of-life sensors to the market.

1. Introduction

Firefighters and other first responders are exposed to a range of hazards while at work, for instance, thermal, chemical, biological, mechanical and nuclear hazards [1]. Furthermore, firefighters are more at risk of exposure to open flame, radiant heat, steam, hot fluids and hot surfaces. Fire-protective clothing contributes to the safety of firefighters by protecting the wearer against heat and flame threats [2]. For this reason, fire-protective clothing must maintain its performance throughout its lifetime.

However, fire-protective clothing can be affected by many ageing conditions, such as heat, UV light, moisture, and abrasion, along with laundering procedures [3]. The mechanical performance of the fire-protective clothing may be reduced due to ageing [4–5-6]. Yet, the loss in performance as a result of ageing can reach a dangerous level before any visual changes are detectable via the human eye [7]. This raises important concerns regarding the firefighters' safety, as well as the safety of workers who wear protective clothing.

At the University of Alberta, in partnership with five industry partners including DTS, as well as the financial support of NSERC, graphene-based end-of-life sensors (heat, UV light, and moisture) are being developed to address this issue [8]. Each sensor consists of several layers, including a conductive track fixed to a sacrificial polymer that is encapsulated in a polymer film. Additionally, each sensor is laid on a textile substrate which will be attached as a patch to specific areas of the outer shell fabric of the firefighter's protective garment.

The sacrificial polymers behave similarly to the outer shell fabrics of firefighters' protective garments when exposed to the ageing conditions of heat, UV light, and moisture [9–10]. The exposure to ageing conditions will initiate a degradation in the sacrificial polymer layer of the sensor, ultimately disrupting the conductivity of the conductive track [8]. The changes in the electrical conductivity will be measured through contacts located on the surface of the sensor using a multimeter during regular protective gear assessment, determining if the fire-protective clothing is safe for further use or not.

2. Manufacturing process

The fabrication of the graphene-based end-of-life sensors includes weaving, treating the textile substrate, putting the sensors' elements together as well as mounting the sensor to the textile substrate. DTS encompasses the production capacity to fabricate these sensors: weaving, finishing, conductive track application, fusing, and product assembly.

2.1. Weaving

The textile substrate is the key component for the development of graphene-based end-of-life sensors. A comprehensive research study was conducted at the University of Alberta to assess and determine the best choice for a textile substrate for the sensors [11]. The purpose of the assessment was to identify a fabric that can resist ageing conditions that are known to be damaging to fire-protective clothing and retain most of its mechanical strength. Moreover, this assessment helped to ensure that the textile substrate does not interfere with the operation of the sensor throughout the life of the garment.

Upon extensive investigation of a wide range of FR fabrics, it was found that Nomex® IIIA, which is a blend of 93% meta-aramid, 5% para-aramid and 2% anti-static fibres, is optimal for the sensor textile substrate as it retains most of its mechanical strength following the exposure to accelerated thermal and hydrothermal ageing, as well as accelerated laundering. DTS utilizes narrow-width looms to manufacture fabrics for highly visible reflective trims (Figure 1). These looms will be utilized for producing the Nomex® IIIA textile substrate for the sensor patches.

2.2. Finishing

Results have shown that Nomex® IIIA demonstrates excellent resistance against heat, moisture, and laundering procedures; however, the fabric is sensitive to UV light, resulting in reduction in mechanical strength [11]. It is necessary that the textile substrate retains its mechanical strength and does not fail before the sensor elements. Hence, surface treatments, such as UV light blockers and absorbers, are being investigated to protect the textile substrate from UV light degradation. DTS currently applies surface treatments for dimensional stability, increased hand, and UV light stabilization using the finishing unit (Figure 2). The finishing unit will be utilized to apply surface treatments to the textile substrate for UV light stability.

2.3. Conductive track

The conductive tracks will be applied as a layer on the sacrificial polymers using different methodologies of application for each sensor [12]. DTS recently purchased a laser scriber, which will be utilized for preparing the conductive tracks.

2.4. Fusing

To protect the sensor components from normal use conditions, such as abrasion and laundering procedures, they will be encapsulated with a polymer film [12]. DTS uses fusing machines to adhere reflective tapes to woven fabrics (Figure 3); these machines will be used to encapsulate the sensor elements, as well as mount the sensor to the textile substrate.

IOP Conf. Series: Materials Science and Engineering 1266 (2023) 012014 doi:10.1088/1757-899X/1266/1/012014

Figure 1. Narrow width loom at DTS.

Figure 2. The finishing unit at DTS.

Figure 3. Narrow fabric and reflective trim entering the fusing machine at DTS.

2.5. Product assembly

DTS is committed to innovation and automation. The goal is to manufacture a continuous strip of sensors, laminate the strip to a continuous band of Nomex® IIIA substrate, and then slit the composite trim down to individual sensors. This production technique will allow for larger capacity production, leading to commercialization.

3. Discussion

Prototypes will be produced at DTS over the next months, and field trials will be completed by the University of Alberta's research team. Once the sensors are deemed viable, DTS will present the sensors to the market, completing the final step of commercialization.

4. Conclusion

Fire-protective clothing is fundamental for the safety of firefighters and is considered the first line of defense against hazardous elements. Hence, monitoring the condition of the fire-protective clothing will help determine its lifetime and verify if it is safe to use or not. A simple useful tool to achieve this goal is the graphene-based end-of-life sensors.

The successful development and commercialization of the sensors will take place at DTS in cooperation with academic researchers from the University of Alberta. Introducing the graphene-based end-of-life sensors to the competitive market of wearable technologies will enhance the safety of firefighters and other users who rely on fire-protective clothing for their wellbeing.

Acknowledgment: This project has received funding from the Natural Sciences and Engineering Research Council of Canada (NSERC) (STPGP 521866 - 18). The authors also want to acknowledge the contribution of project team members: Dr. C. Cho, Dr. C. Braun, A. Saha, Md. S. Hoque, L. Zheng, R. Khalkhali, Z. Lin, M. Khemir, N. Samaraweera, Dr. A. Chhetry, A. Pimenta Bonatto, M. Osama, and J. Forgie.

References

[1] Song G, Mandal S and Rossi R M 2017 1 - Introduction *Thermal Protective Clothing for Firefighters* (Woodhead Publishing) pp 1–4 (https://doi.org/10.1016/B978-0-08-101285-7.00001-0)

[2] McQuerry M, Klausing S, Cotterill D and Easter E 2015 A Post-use Evaluation of Turnout Gear Using NFPA 1971 Standard on Protective Ensembles for Structural Fire Fighting and NFPA 1851 on Selection, Care and Maintenance *Fire Technology* **51** 1149–66

[3] Rezazadeh M and Torvi D A 2011 Assessment of Factors Affecting the Continuing Performance of Firefighters' Protective Clothing: A Literature Review *Fire Technol* **47** 565–99

[4] Dolez P I, Tomer N S and Malajati Y 2019 A quantitative method to compare the effect of thermal aging on the mechanical performance of fire protective fabrics *Journal of Applied Polymer Science* **136** 47045

[5] Arrieta C, David É, Dolez P and Vu-Khanh T 2011 Hydrolytic and photochemical aging studies of a Kevlar®-PBI blend *Polymer Degradation and Stability* **96** 1411–9

[6] Dolez P and Malajati Y 2020 Resistance of Fire Protective Fabrics to Repeated Launderings (West Conshohocken: ASTM International) pp 100-113 (https://doi.org/10.1520/STP162420190079)

[7] Rossi R M, Bolli W and Stämpfli R 2008 Performance of firefighters' protective clothing after heat exposure *Int J Occup Saf Ergon* **14** 55–60

[8] Dolez P, Chung H-J and Cho C 2022 End-of-Life Sensors for Fabrics. PCT International Patent Application PCT/CA2022/000006. 2022/03/03

[9] Cho C, Nam S L, de la Mata A P, Harynuk J J, Elias A L, Chung H-J and Dolez P I 2022 Investigation of the accelerated thermal aging behavior of polyetherimide and lifetime prediction at elevated temperature *Journal of Applied Polymer Science* **139** 51955

[10] Braun C A, Nam S L, de la Mata A P, Harynuk J, Chung H-J and Dolez P I 2022 Hydrothermal aging of polyimide film *Journal of Applied Polymer Science* **139** 52183

[11] Yehia D 2021 *Investigation of Support Fabrics for Graphene-Based End-of-Life Sensors for Fire Protective Garments* Master Thesis, University of Alberta, Canada

[12] Cho C 2021 *Investigation of polyetherimide's thermal degradation and fabrication of thermal end-of-life sensor for fire protective clothing* PhD Thesis, University of Alberta, Canada

Green and sustainable dyeing of wool with madder plant of the Moroccan flora

F E Mesrar[1], H Tachallait[1], O Cherkaoui[2], K Bougrin[1,3] and R Benhida[1,4*]
[1] Department of Chemical and Biochemical Sciences, Green Process Engineering (CBS), Mohammed VI Polytechnic University (UM6P), Benguerir, Morocco
[2] Laboratoire de recherche sur les matériaux textiles (REMTEX), Ecole Supérieure des Industries du Textiles et de l'Habillement (ESITH), Casablanca, Morocco
[3] Equipe de Chimie des Plantes et de Synthèse Organique et Bioorganique, URAC23, Faculty of Science, B.P. 1014, Geophysics, Natural Patrimony and Green Chemistry (GEOPAC) Research Center, Mohammed V University, Rabat, Morocco
[4] Institut de Chimie de Nice (ICN), UMR 7272, Université Côte d'Azur, CNRS, Nice, France
E mail : Rachid.benhida@um6p.ma

Abstract. Textile industry is evolving towards limiting the use of synthetic dyes due to their environmental and health concerns. Natural dyes are considered the ultimate alternative as it is green and ecofriendly. However, natural dyeing is limited by the use of metal mordants to enhance the colorimetric and fastness behavior of dyed textiles. In this work, madder roots (Rubia Tinctorum) from Morocco were analyzed by mean of HPLC-PDA and used to dye wool yarns. The effect of pre-mordanting with different metal mordants was assessed. Color properties, exhaustion, and fixation rates of dyed wool was also evaluated depending on the mordants.

1. Introduction

Up to date, textile industry has been mainly based on the use of synthetic dyes and colorant due to their high color fastness, dyeing reproducibility and cheaper prices [1,2]. However, in the last few years, synthetic dyes were strongly cited as toxic and dangerous, causing irritation or dermatitis, and several environmental problems [3–6]. Consequently, due to the increased environmental awareness clothing industries were forced to reduce its use of toxic materials and to use more suitable and ecological dyes [7].

Natural dyes are colorants derived from nature and do not need any chemical processing [8–10]. Since antiquity, natural dyes have been widely used in human everyday life in a variety of fields, including in the textile industry. They are ecological, safe, non-toxic, non-carcinogenic and biodegradable [11,12]. In addition, natural dyes exhibit several functional properties like antimicrobial activity [13], UV-protection [14] , and insect repellent [15].

Nevertheless, natural dyes have also some drawbacks limiting their extended uses such as poor fastness, difficulties in color reproducibility, lack of standardization methods, and process complexity [16]. To overcome these limitations, mordanting is considered the most complex and limiting step in natural dyeing [17]. Mordants are substances that form a complex between fiber and dye molecules and therefore, enhance the color stability and the fastness properties of naturally dyed materials [18,19].

IOP Conf. Series: Materials Science and Engineering 1266 (2023) 012015 doi:10.1088/1757-899X/1266/1/012015

Madder roots represent a very famous source of vegetable red dye for textile natural dyeing. *Rubia tinctorum* L. (Madder) is a herbaceous perennial plant that is growing naturally in Europe, and Mediterranean countries [20]. The main dye components present in madder roots have an anthraquinone skeleton in the glycosidic and aglycon forms [22] (Table 1). The composition of madder depends on several factors such as the geographical location and the growth conditions, yet the most best-known anthraquinone derived from *Rubia tinctorum* is alizarin [22,23].

In order to get further insight into the mordanting step with natural products, we investigate in this work the colorimetric properties of wool fabric dyed with madder roots (Rubia Tinctorum) from Morocco by using different metal mordants.

Table 1. Anthraquinone structures.

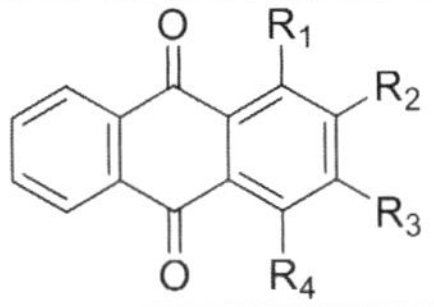

Name	**R_1**	**R_2**	**R_3**	**R_4**
Lucidin primeveroside	OH	CH_2OH	O-primeveroside	H
Ruberythric acid	OH	O-primeveroside	H	H
Alizarin	OH	OH	H	H
Purpurin	OH	OH	H	OH

2. Material and Methods

2.1. Materials

Wool fabric ready to dye was obtained from a textile mill in Casablanca, Morocco. Powdered madder roots (*Rubia Tinctorum*) were purchased from local herbal market in Marrakech, Morocco. The mordants used: alum, Copper (II) sulphate, Iron (II) sulphate, Magnesium sulphate, Calcium sulphate, and Manganese sulphate were of commercial grades. Water, Acetonitrile, and Formic acid were of HPLC grades and purchased from Merck.

2.2. Chromatographic analysis

2.5 g of powdered madder roots was refluxed at 70°C for 3 hours in 100 ml of Ethanol-water (75-25). The obtained extract was filtered over a 0.45 µm membrane filter, diluted in methanol-water (1-1), and injected in HPLC-DAD. Chromatographic analysis of *Rubia Tinctorum* and the identification of the resulting peaks was determined through comparison with the results described by [25]

Chromatographic analysis was carried out using JASCO LC-4000 Series (Jasco, France) equipped with an AS-4150 autosampler, PU-4180 pump, CO-4061 oven, and MO-4010 photo diode array detector. An agilent Zorbax SB-C8 analytical column (4.6×250 mm, 5 µm) was used. PDA detection was performed by scanning from 200 to 650 nm, and the chromatographic peaks were plotted at 250 nm. The data were treated with the ChromNav software by Jasco. The separation was carried out using two solvents: (A) water + 0,1% formic acid; (B) acetonitrile + 0,1% formic acid with a flow rate of 0.7 mL/min and following a linear elution program (Table 2).

Table 2: Gradient programme for HPLC analysis.

Time (min)	Water + 0,1% formic acid	Acetonitrile + 0,1% formic acid
7	70	30
10	30	70
20	30	70
25	0	100
35	0	100
40	70	30
45	70	30

2.3. Dyeing procedure

Dye solution was prepared by heating dried and powdered madder roots in water (liquor ratio 1/50) at 77°C for 1 hour. The obtained dye solution was filtered and used for dyeing.

Previously rinsed wool was mordanted with pre-mordanting method. Wool yarns were immersed in aqueous solutions of 2%, 5% and 10% (o.w.f) of alum, Copper (II) sulphate, Iron (II) sulphate, Magnesium sulphate, Calcium sulphate, and Manganese sulphate keeping liquor ratio 1/50. The mordanting bath was heated at 67°C for 1 hour, then the samples were washed with water and air dried to remove the excess mordant remained on wool surface.

Mordanted wool yarns were placed in dye bath keeping liquor ratio of 1/50 and heated at 76°C for 1 hour. The dyed wool was then washed thoroughly with cold water, squeezed and air dried.
To examine the efficiency of each mordant, exhaustion (%E) and fixation (%F) rates were determined. The exhaustion and fixation rates are calculated 'as in equation (1) and (2)' respectively:

$$\%E = \frac{E_a - E_b}{E_a}\,100 \qquad (1)$$

$$\%F = \frac{E_a - E_b - E_c}{E_a}.100 \qquad (2)$$

Where:
Ea: absorbance of the initial dye solution
Eb: absorbance of the final dye solution
Ec: absorbance of the after washing solution

2.4. Color measurement

CIELab coordinates and color strength (K/S) were determined with a Data Color 600TM spectrophotometer with a D65 illuminant and a 10° standard observer. L*, a* and b* represent lightness, redness–greenness and yellowness–blueness of color, respectively. Color strength values (K/S) of dyed wool yarns were also calculated from diffuse reflectance at λ according to Kubelka–Munk 'as in equation (3)'.

$$\frac{K}{S} = \frac{(1-R)^2}{2R} \qquad (3)$$

where R is reflectance at λ, K is absorption coefficient and S is scattering coefficient.

3. Results and discussion

3.1. HPLC analysis

For a qualitative analysis and an elucidation of its composition, we have conducted the HPLC-PDA analysis of Moroccan Madder (Rubia Tinctorum). The obtained profile is depicted in 'Figure 2'. Water-ethanol (1-1) profile showed 9 main peaks. As an initiatory investigation, we built our chromatographic study on the previous work of Derksen et al.[25]. Consequently, we were able to estimate the nature of the major peaks. According to the [25], the peak 1 is Lucidin primeveroside, peak 2 is Ruberythric acid, peak 3 is Alizarin, and peak 4 is Purpurin.

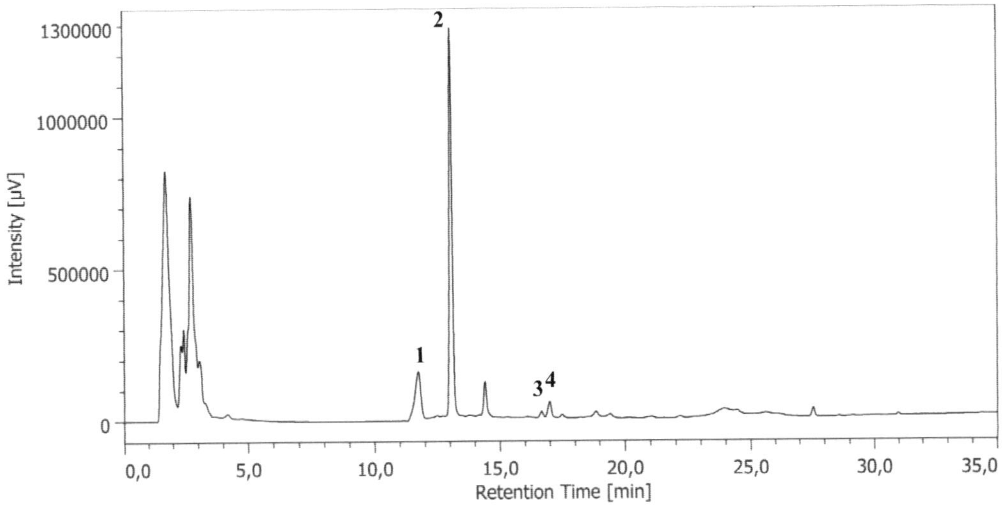

Figure 1.HPLC profile of ethanol-water (1-1) extract of madder roots (*Rubia Tinctorum*).

3.2. Dyeing method

Table 3: Colorimetric parameters of wool fabric dyed with madder roots.

Mordants	Concentration (%owf)	Exhaustion rate (%)	Fixation rate (%)	L*	a*	b*	K/S	Image of dyed wool
No mordant	-	33,33	29,77	33,87	50,08	32,33	1,59	
Alum	2%	50,15	48,17	31,45	39,6	25,18	1,98	
	5%	53,17	50,56	29,48	37	22,64	2,52	
	10%	57,10	54,42	16,35	10,22	4,03	3,13	
$FeSO_4$	2%	58,68	57,32	15,23	6,39	1,62	16,02	
	5%	52,5	50,62	15,4	5,41	0,27	20,94	
	10%	51,26	49,60	26,71	22,6	11,57	22,05	
$CuSO_4$	2%	35	33,51	26,55	21,11	11,16	5,75	
	5%	35,62	32,96	32,11	21,38	11,23	5,89	
	10%	38,01	35,32	25,15	33,04	19,9	3,71	
$MnSO_4$	2%	45,89	44,39	28,48	36,7	24,5	4,44	
	5%	42,67	40,16	25,94	35,09	22,31	3,21	
	10%	41,49	39,32	27,46	38,12	24,91	4,01	
$CaSO_4$	2%	36,32	34,11	24,59	35,38	24,4	3,43	
	5%	40,04	36,55	27	35,38	23,67	4,35	
	10%	42	38,18	32,87	38,4	24,81	3,54	
$MgSO_4$	2%	48,32	45,75	28,05	35,02	22,49	2,14	
	5%	46,41	45,01	28,1	36,81	23,69	3,39	
	10%	40,25	38,59	33,87	50,08	32,33	3,27	

Wool yarns were dyed with madder and pre-mordanted with different metal salts as mentioned in Table 4. Mordanting has improved the color coordinates, the exhaustion and the fixation rates of the dyed wool compared to un-mordanted fabric. According to the results, un-mordanted wool have the lowest color strength, exhaustion and fixation rates compared to mordanted wool. Metal mordants have the capability to create coordination complexes with dye molecules and the fabric at the same time [26]. The dye molecules are thus strongly bound to the substrate resulting in higher exhaustion and color strength. Among the mordants used 10% $FeSO_4$ showed the best color strength and the lowest L* value resulting in a darkening effect on madder dyed wool. It is known that iron salts have strong coordination tendency which lead to the formation of a complex with the fiber and the dye [27–31]. Alum mordant on the other hand, had low K/S values when compared to the other mordants. Alum tends to coordinate with the dye rather than the fiber resulting in less interaction and lower color strength [29].

As can be seen from Table 4, color strength follows the order: un-mordanted < 10% Alum < 5% $MgSO_4$ < 5% $CaSO_4$ < 2% $MnSO_4$ < 5% $CuSO_4$ < 10% $FeSO_4$.

CieLab experiments clearly show that the color coordinates of all the dyed samples are in the red-yellow quadrant of the CIEL*a*b* color space. Mordants also changed the shades of dyed fabrics when compared to samples dyed without mordant. Mordanted and dyed fabrics showed vivid and strong color, whereas fabrics dyed without mordant showed a light shade. In addition, Alum alters the color shades to the yellow region of the color space, whereas ferrous sulfate shifts the color coordinates to the red part. It can be concluded that the type of mordants had a significant impact on the color properties of the dyed fabrics.

4. Conclusion

This work focuses on studying the effect of different mordants on the dyability of wool yarns with madder (*Rubia Tinctorum* L.) from Morocco. It was found that mordanting can dramatically improve the colorimetric properties and dye uptake of wool fabrics. In addition, HPLC-PDA analysis of madder extract was provided. Further work will aim to purify and quantify the composition of Moroccan madder roots and to study its dyeing behaviour on natural fibers.

Acknowledgments

The authors would like to express appreciation for the support of the sponsors [OCP group, Mohammed VI Polytechnic University, Ecole Supérieure des Industries du Textile et Habillements (ESITH)].

References

[1] Nidheesh, P. V., Zhou, M., and Oturan, M. A., 2018, "An Overview on the Removal of Synthetic Dyes from Water by Electrochemical Advanced Oxidation Processes," Chemosphere, **197**, pp. 210–227.

[2] Nambela, L., Haule, L. V., and Mgani, Q., 2020, "A Review on Source, Chemistry, Green Synthesis and Application of Textile Colorants," J. Clean. Prod., **246**.

[3] Uddin, F., 2021, "Environmental Hazard in Textile Dyeing Wastewater from Local Textile Industry," Cellulose, **28**(17), pp. 10715–10739.

[4] Samsami, S., Mohamadi, M., Sarrafzadeh, M. H., Rene, E. R., and Firoozbahr, M., 2020, "Recent Advances in the Treatment of Dye-Containing Wastewater from Textile Industries: Overview and Perspectives," Process Saf. Environ. Prot., **143**, pp. 138–163.

[5] Yusuf, M., 2019, "Synthetic Dyes: A Threat to the Environment and Water Ecosystem," Text. Cloth. Environ. Concerns Solut., pp. 11–26.

[6] Ali, H., 2010, "Biodegradation of Synthetic Dyes - A Review," Water. Air. Soil Pollut., **213**(1–4), pp. 251–273.

[7] Yusuf, M., Shabbir, M., and Mohammad, F., 2017, "Natural Colorants: Historical, Processing and Sustainable Prospects," Nat. Products Bioprospect., **7**(1), pp. 123–145.

[8] Khan, A. A., Iqbal, N., Adeel, S., Azeem, M., Batool, F., and Bhatti, I. A., 2014, "Extraction of Natural Dye from Red Calico Leaves: Gamma Ray Assisted Improvements in Colour Strength and Fastness Properties," Dye. Pigment., **103**, pp. 50–54.

[9] Yusuf, M., Shahid-Ul-Islam, Khan, M. A., and Mohammad, F., 2016, "Investigations of the

Colourimetric and Fastness Properties of Wool Dyed with Colorants Extracted from Indian Madder Using Reflectance Spectroscopy," Optik (Stuttg)., **127**(15), pp. 6087–6093.

[10] Adeel, S., Hussaan, M., Rehman, F. ur, Habib, N., Salman, M., Naz, S., Amin, N., and Akhtar, N., 2019, "Microwave-Assisted Sustainable Dyeing of Wool Fabric Using Cochineal-Based Carminic Acid as Natural Colorant," J. Nat. Fibers, **16**(7), pp. 1026–1034.

[11] Kumar Samanta, A., and Agarwal, P., 2009, "Application of Natural Dyes on Textiles," Indian J. Fibre Text. Res., **34**, pp. 384–399.

[12] Shahid, M., Shahid-Ul-Islam, and Mohammad, F., 2013, "Recent Advancements in Natural Dye Applications: A Review," J. Clean. Prod., **53**, pp. 310–331.

[13] Kamboj, A., Jose, S., and Singh, A., 2021, "Antimicrobial Activity of Natural Dyes–A Comprehensive Review," J. Nat. Fibers.

[14] Hou, X., Chen, X., Cheng, Y., Xu, H., Chen, L., and Yang, Y., 2013, "Dyeing and UV-Protection Properties of Water Extracts from Orange Peel," J. Clean. Prod., **52**, pp. 410–419.

[15] Shakyawar, D. B., Raja, A. S. M., Kumar, A., and Pareek, P. K., 2015, "Antimoth Finishing Treatment for Woollens Using Tannin Containing Natural Dyes," Indian J. Fibre Text. Res., **40**(2), pp. 200–202.

[16] Patel, B. H., 2011, "Natural Dyes," Handb. Text. Ind. Dye. Princ. Process. Types Dye., **1**, pp. 395–424.

[17] Manian, A. P., Paul, R., and Bechtold, T., 2016, "Metal Mordanting in Dyeing with Natural Colourants," Color. Technol., **132**(2), pp. 107–113.

[18] İşmal, Ö. E., and Yildirim, L., 2018, "Metal Mordants and Biomordants," *The Impact and Prospects of Green Chemistry for Textile Technology*, Elsevier, pp. 57–82.

[19] Hosseinnezhad, M., Gharanjig, K., Jafari, R., and Imani, H., 2021, "Green Dyeing of Woolen Yarns with Weld and Madder Natural Dyes in the Presences of Biomordant," Prog. Color. Color. Coatings, **14**(1), pp. 35–45.

[20] "Encyclopedia of Applied Plant Sciences | ScienceDirect" [Online]. Available: https://www.sciencedirect.com/referencework/9780123948083/encyclopedia-of-applied-plant-sciences. [Accessed: 15-Apr-2022].

[21] De Santis, D., and Moresi, M., 2007, "Production of Alizarin Extracts from Rubia Tinctorum and Assessment of Their Dyeing Properties," Ind. Crops Prod., **26**(2), pp. 151–162.

[22] Cuoco, G., Mathe, C., Archier, P., Chemat, F., and Vieillescazes, C., 2009, "A Multivariate Study of the Performance of an Ultrasound-Assisted Madder Dyes Extraction and Characterization by Liquid Chromatography-Photodiode Array Detection," Ultrason. Sonochem., **16**(1), pp. 75–82.

[23] Mehrparvar, L., Safapour, S., Sadeghi-Kiakhani, M., and Gharanjig, K., 2016, "A Cleaner and Eco-Benign Process for Wool Dyeing with Madder, Rubia Tinctorum L., Root Natural Dye," Int. J. Environ. Sci. Technol., **13**(11), pp. 2569–2578.

[24] Ford, L., Rayner, C. M., and Blackburn, R. S., 2018, "Comparative Sorption Isotherms for Colorants Present in Dyers' Madder (Rubia Tinctorum L.) Provide New Insights into Historical Dyeing," Color. Technol., **134**(1), pp. 3–12.

[25] Derksen, G. C. H., Van Beek, T. A., De Groot, Æ., and Capelle, A., 1998, "High-Performance Liquid Chromatographic Method for the Analysis of Anthraquinone Glycosides and Aglycones in Madder Root (Rubia Tinctorum L.)," J. Chromatogr. A, **816**(2), pp. 277–281.

[26] Ghaheh, F. S., Nateri, A. S., Mortazavi, S. M., Abedi, D., and Mokhtari, J., 2012, "The Effect of Mordant Salts on Antibacterial Activity of Wool Fabric Dyed with Pomegranate and Walnut Shell Extracts," Color. Technol., **128**(6), pp. 473–478.

[27] Uddin, M. G., 2014, "Effects of Different Mordants on Silk Fabric Dyed with Onion Outer Skin Extracts," J. Text., **2014**, pp. 1–8.

[28] Ghaheh, F. S., Nateri, A. S., Mortazavi, S. M., Abedi, D., and Mokhtari, J., 2012, "The Effect of Mordant Salts on Antibacterial Activity of Wool Fabric Dyed with Pomegranate and Walnut Shell Extracts," Color. Technol., **128**(6), pp. 473–478.

[29] Bhattacharya, S. D., and Shah, A. K., 2000, "Metal Ion Effect on Dyeing of Wool Fabric with Catechu," Color. Technol., **116**(1), pp. 10–12.

[30] Moiz, A., Aleem Ahmed, M., Kausar, N., Ahmed, K., and Sohail, M., 2010, "Study the Effect of Metal Ion on Wool Fabric Dyeing with Tea as Natural Dye," J. Saudi Chem. Soc., **14**(1), pp. 69–76.

[31] Shahmoradi Ghaheh, F., Moghaddam, M. K., and Tehrani, M., 2021, "Comparison of the Effect of Metal Mordants and Bio-Mordants on the Colorimetric and Antibacterial Properties of Natural Dyes on Cotton Fabric," Color. Technol., **137**(6), pp. 689–698.

Study and characterization of Moroccan wool fibers for the development of non-woven products

Nabil Aazou[1,2,3], Ayoub Nadi[1], Omar Cherkaoui[1], Said Gmouh[2]
[1] REMTEX Laboratory, Higher School of Textile and Clothing Industries (ESITH), Km 8, Route d'El Jadida, Casablanca, Morocco.
(E-mail: aazou@esith.ac.ma ; nadi@esith.ac.ma ; cherkaoui@esith.ac.ma)
[2] LIMAT Laboratory, Faculty of Sciences Ben M'Sik, Hassan II University of Casablanca, B. P. 7955, Casablanca, Morocco.
(E-mail: said.gmouh@gmail.com)
[3] Author to whom any correspondence should be addressed

aazou@esith.ac.ma

Abstract. Wool fibers have been used as textile raw materials for clothing and home textiles for many years. Recently, it has also been increasingly used in technical textile applications. However, man-made fibers dominate the market and minimize the importance of wool in the textile industry. On the other hand, the great future opportunity for wool is the growing interest of consumers in natural, renewable and sustainable materials. In addition, in the past few decades, wool production has improved significantly in terms of yield and quality.
In this study we have characterized and applied different analysis on the physical and chemical properties of raw Moroccan wool using various instruments such as FTIR and OFDA, etc., and several test methods according international standards like grease rate, alkali content, acid content, Solubility in alkali and Tensile strength, etc. The aim of our study is to determine the potential use of wool for elaboration of nonwoven textile for a technical use and determine their end-use applications.
Keywords: Natural fiber; Moroccan wool; Nonwoven; Technical textile.

1. Introduction:

Textile processing industry is characterized not only by the large volume of water required for various unit operations, but also by the variety of chemicals used for various processes. There is a long sequence of wet processing stages requiring input of water, chemical and energy thus generating wastes at each stage including liquid, gaseous and solid wastes, which may be hazardous [1]. Therefore, Scientists are constantly looking for alternative solutions to reduce their release into nature.

Natural fibers include fibers obtained from animal or vegetable sources without any chemical treatment. This class of fibers is best known for being environmentally friendly, biodegradable, low toxicity and less allergenic than synthetic fibers [2]. The wool fiber from the various breeds of sheep is by far the most commonly-used animal fiber [3, 4]. Wool fiber may be a sustainable and renewable carbon fiber precursor. The high-value application of waste wool will also alleviate the current disposal of wool waste, thereby improving our environment.

However, in Morocco, the wool fibers are considered as wasted and with natural resource. For this reason, this natural resource needs to be valorized for the diverse advantages and applications the human being can get from it.

Our research follows this context and aims at studying and characterizing the Moroccan raw wool from different provenance for the elaboration of non-woven textile. Our work is composed of three parts that are:

- The study of the chemical properties in terms of Grease rate, water absorption capacity, Moisture, Acid and Alkali percentage, etc.
- The study of the physical properties with regard to the fiber diameter and the tensile strength.
- The reach of an appropriate application based on the results found on the previous parts.

2. Material and Methods:

2.1. Products and Solvents:

All chemicals and reagents were used as received, without further purification:
Dichloromethane and Sodium Hydroxide from SIGMA-ALDRICH; Pyridine and Boric acid from LOBACHEMIE; Phenolphthalein and Methylene Blue from FLUKA; Acetic Acid and Ammonia from SOLVACHIM; Hydrochloric acid from HONEYWELL; Methyl Red from MERCK and ECE Phosphate Reference Detergent 'B' from SDC Enterprises.

2.2. Sampling:

The wool fibers used in this study were collected from different regions of Morocco as shown in the table 1.

Table 1: Provenances and backgrounds of Moroccan wool samples

Race	Timahdite	Bni Guil	Sardi	Boujaad	D'man	Black of Siroua	White of Siroua	Saghro
Abbreviation	T	BG	S	B	D	B.S	W.M	SG
Provenance	Ifran	Oujda	Settat	Bejaad	Zagoura	Taroudant	Taroudant	Tinghir

The purpose of the following tests is to find out the appropriate application for these fibers.

2.3. Study of physicochemical properties of wool fibres:

The physicochemical properties of wool fibers have been studied in terms of Grease Rate, water absorption capacity, Moisture, Acid and Alkali percentage and their Solubility in alkali. These tests were determined according to their appropriate standards.

2.4. Characterization of wool fibres:

The chemical structure and functional groups of the raw wool fibres were studied by Fourier Transformed Infrared Spectroscopy using FTIR Spectrometer Thermo Scientific Nicolet is20 by recording transmission mode analyses (%) in the 4000-500 cm^{-1} range. The measurement of the diameter of wool fibers was investigated to determine their fineness using Ofda Fibre Fineness Meter with a magnification of x0.25. The tensile strength and elongation measurements of the fibres was measured with the help of a 5N cell and a speed of 10 mm/min by the LUDWIG Mpk tensile testing machine.

3. Results and discussion:

3.1. Study of chemical properties of Moroccan wool fibres:

The FTIR Spectroscopy analysis was investigated to determine the functional groups of the different samples of Moroccan wool (figure 1). All the samples showed similar spectra denoted by a band 3272-3072 cm^{-1} with a wide and intense peak indicating the presence of vibrations of elongations of O-H and N-H bonds associated, then we noticed a band at 1632 cm^{-1} of only one peak of weak intensity corresponding to elongations of the Primary amide bonds, and a very fine bands 1523 cm^{-1} of medium intensity indicating to the presence of deformations in the plane of the Secondary Amide bonds, and finally a bands between 1442 and 1380 cm^{-1} with a very fine and single peak of low intensity, corresponding to the presence of vibrations deformations of Tertiary Amide. After comparing the characteristic bonds of the raw samples, we find that they have the same structural composition and the only difference between them is the variation of the intensities of the functional groups and we can confirm this by the diversity of the sizes and diameters of the fibers. Similar results of infrared spectrum were observed by Yao et al **[5]**.

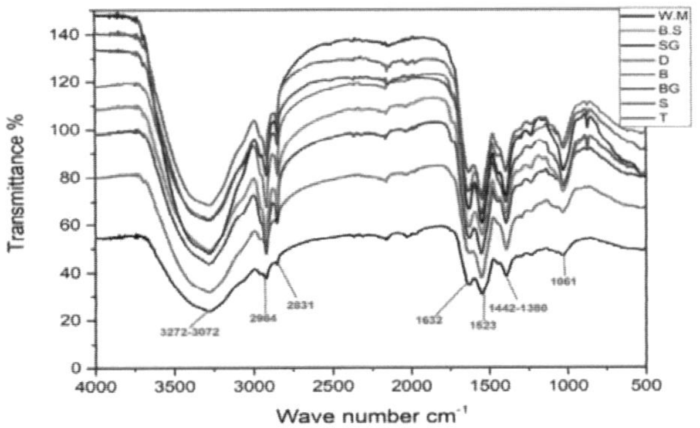

Figure 1: FTIR spectrum of different Moroccan wool fibers

The table 2 shows the physico-chemical properties of the different breeds studied. For the percentage of fatty acids, the extract of the (S) breed contains a higher proportion of fatty acid substances than the other breeds with a value of 12.12% and also absorbs more moisture than the other breeds which are adequate with the results found by *Sengupta et al* **[6]**.

Table 2: Chemical Properties of wool fibers:

	Physicochemical properties					
Standard	ISO3074-2014	ISO 3073-1975	ISO 2916-1975	ASTM D1576-13	ASTM D570	ISO 3072-1975
Breeds	***Fat content (%)***	***Acid content (%)***	***alkali content (%)***	***Moisture (%)***	***Water absorption capacity (%)***	***Solubility in alkali (%)***
S	12,12	0,341	1,647	10,142	350	17,655
T	4,67	0,581	2,546	7,296	406	14,934
BG	3,32	0,194	1,915	7,023	288	17,766
B	8,79	0,161	2,702	9,414	272	10,565
D	8,59	0,274	1,713	6,372	208	15,507
SG	9,49	0,41	2,045	4,343	406	18,405
W.M	2,38	0,232	1,855	9,199	396	16,79
B.S	7,3	0,343	1,46	7,147	314	13,747

On the other hand, the acid content test shows that the acid contained within the wool fibers for all the Moroccan breeds does not exceed 0.6% of the dry mass of the samples studied, and the (T) breed contains the highest amount of acid with a value of 0.581%. However, for the alkali content, shows that the alkali contained in the wool fibres for the different breeds is also lower and does not exceed 2.8% of the dry mass of the samples studied, and the (B) breed contains the highest amount of alkali with a value of 2.702% and it is also the most resistant to the attack by alkaline solutions as shown from the solubility test.

The test of water absorption capacity by the 8 breeds is also performed. We verified that the **T** and **SG** breeds have the highest water absorption power which is explained by the increase of the initial mass up to 406%.
These results can provide advantages in different aspects for example the application in the dyeing industry it gives us an idea about what breed can retain dyestuffs very well.

3.2. Study of physical properties of Moroccan wool fibres:

We compared in the table 3 the mean diameters of each race and we can conclude that the race **D** has the finest fibres with a value of 25.54 µm and a C.V of 2.065%, on the other hand the average diameter of the race **B.S** is the biggest compared to the other races with a value of 46.959 µm and a C.V of 2.811%.

Table 3: Mean Diameter of the Moroccan wool fibers

Breeds	T	BG	S	B	D	B.S	W.M	SG
Mean Diameter:	29,742	41,20	40,230	28,421	25,54	46,959	44,98	30,968
Coefficient of variation C.V:	5,27%	5,01%	4,25%	3,297%	2,065%	2,811%	2,612%	3,063%

From the figure 2 and the table 4, it can be deduced that the B.S breed has the highest elongation which is equal to 39.87% with average stress at break and quite high force at break. However, the SG breed has the lowest elongation among the breeds with average values for stress and force at break. Similar results were observed by *Moody et al* **[7]**.

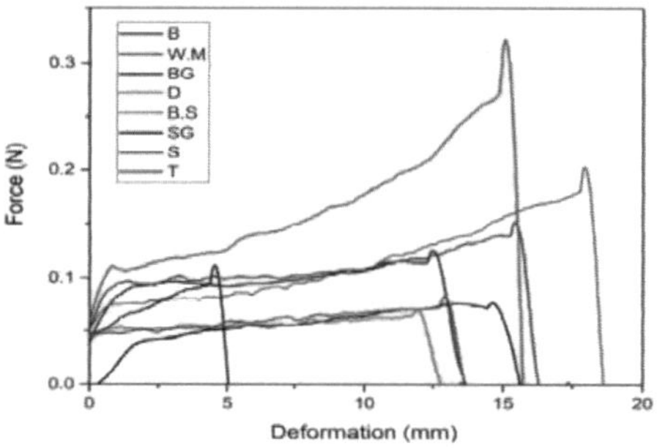

Figure 2: Tensile strength plot of the samples

Table 4: Tensile properties of the samples

Breeds:	Elongation (%):	Stress (MPa):	Force (N) :
T	34.24	218.833	0.152
S	28.71	65.713	0.084
BG	24.93	34.619	0.115
B	29.67	120,327	0.076
B.S	39.87	117.234	0.203
W.M	33.47	64.375	0.321
D	26.49	136.855	0.070
SG	10.09	148.77	0.112

3.3. Application based on different results

Wool constitutes 5% of the world textile fiber production. Wool's limited supply coupled with its excellent textile performance properties has resulted in wool products being positioned in the luxury end of most markets. The promotional activities of IWS have focused on the "woolmark" and have assisted greatly with enhancing this position in the market place. The "woolmark" is an internationally recognized registered trade mark denoting products made from pure new wool, which also meets strict performance criteria **[8]**.

Owing to the range of sheep breeds and wool types produced wool is processed into an equally wide range of end products. End uses of wool is determined by the main fiber characteristics. These include fiber diameter, fiber length/strength, color and handle, plus others such as luster, bulk, presence of faults like vegetable matter and the level of felting **[8]**.

The two major end uses for wool are apparel products and carpets (both machine made and hand-knotted).

The three systems for making wool yarns are:

- Woolen (condenser spun)
- Semi-worsted, or
- Worsted.

According to the previous results, we can summarize the main end uses of Moroccan wool fibers as illustrated in table 5.

Table 5: Main end uses of Moroccan wool fibers

Name	Timahdite	Bni Guil	Sardi	Boujaad	D'man	Black of Siroua	White of Siroua	Saghro
Mean Diameter	29,742	41,20	40,230	28,421	25,54	46,959	44,98	30,968
Fat content (%)	4,67	3,32	12,12	8,79	8,59	7,3	2,38	9,49
Acid content (%)	0,581	0,194	0,341	0,161	0,274	0,343	0,232	0,41
alkali content (%)	2,546	1,915	1,647	2,702	1,713	1,46	1,855	2,045
Moisture (%)	7,296	7,023	10,142	9,414	6,372	7,147	9,199	4,343
Water absorption (%)	406	288	350	272	208	314	396	406
Solubility in alkali (%)	14,934	17,766	17,655	10,565	15,507	13,747	16,79	18,405
Applications	Woolen Fabrics; Hand Knitting Yarn; Machine Knitting; Blankets; Upholstery; Bedding.	Carpets.	Carpets.	Woolen Fabrics; Hand Knitting Yarn; Machine Knitting; Blankets; Upholstery; Bedding.	Worsted Fabrics; Woolen Fabrics; Hand Knitting Yarn; Machine Knitting.	Carpets.	Carpets.	Hand Knitting Yarn; Blankets; Upholstery; Bedding; Carpets.

There are however, a vast number of other products made from wool and a few of the more interesting ones include:

❒ Mattress and furniture stuffings
❒ Tennis ball covers
❒ Piano hammers
❒ Insulation
❒ Felt hats and toys.

4. Conclusion

This work aims at studying and characterizing the Moroccan wool to elaborate a textile for technical use. For that, in a first step, we studied the wool of all the breeds located in different regions of Morocco. We have found that each breed is intended for a specific application based on its own physico-chemical properties. Thus, through FTIR, OFDA, fat content, solubility and moisture analyses, etc., we have been able to identify and characterize the wool of all Moroccan breeds.

The perspectives are divided into three parts: the first part is related to further analysis of the different results to reach appropriate application of each Moroccan wool breed. The second part is the elaboration of a non-woven textile based on different Moroccan wool breed. The last part concerns the study of the performance in terms of tensile strength, abrasion resistance, air and the water permeability, the flammability and thermal conductivity.

Acknowledgment: This research was supported by ESITH. We thank our colleagues from FS Ben M'Sik who provided insight and expertise that greatly assisted the research.

References:

[1] Turley, D., et al., London, UK, 2009. The role and business case for existing and emerging fibres in sustainable clothing: final report to the Department for Environment, Food and Rural Affairs (Defra).

[2] Erdogan, U. H., Seki, Y., & Selli, F. (2020). Wool fibres. In *Handbook of natural fibres* (pp. 257-278). Woodhead Publishing.

[3] TextileExchange. (2021). *Preferred Fiber & Materials Market Report 2020*. TextileExchange.

[4] Wool Market. (2019). Global Market Study on Wool: persistencemarketresearch.com.

[5] Yao, J., Liu, Y., Yang, S., & Liu, J. (2008). *Characterization of Secondary Structure Transformation of Stretched and Slenderized Wool Fibers with FTIR Spectra.*

[6] Sengupta, A., & Behera, J. (2014). Comprehensive view on chemistry, manufacturing & applications of lanolin extracted from wool pretreatment. *American Journal of Engineering Research*, *3*(7), 33-43.

[7] Moody, V., & Needles, H. L. (2004). 3—Major Fibers and Their Properties. *Tufted Carpet*, 35-59.

[8] Petrie, O. J. (1995). *Harvesting of textile animal fibres* (No. 122). Food & Agriculture Org.

Biodegradation properties of natural fibers for agro textile nonwovens production

R Brunsek[*1], N Jugov[2], P Marasovic[3] and A Mioč[4]

[*1] Associate professor, Department of materials, fibers and textile testing, University of Zagreb Faculty of Textile Technology, Croatia

[2] Project associate, Department of Textile Design and Management, University of Zagreb Faculty of Textile Technology, Croatia

[3] PhD student, Department of Textile Design and Management, University of Zagreb Faculty of Textile Technology, Croatia

[4] Student, University of Zagreb Faculty of Textile Technology, Croatia

E-mail: ruzica.brunsek@ttf.unizg.hr

Abstract. The aim of this paper is to investigate biodegradation properties of PLA (as a bio-based polymer), Hemp, Jute (as natural bast fibers) and Viscose (as a cellulose regenerates) for agrotextile nonwovens production. For the purpose of those investigations, an analysis of the biodegradation time of samples were performed by soil burial test (ISO 11721 – *Textiles – Determination of resistance of cellulose-containing textiles to microorganisms – Soil burial test*) conducted under controlled conditions in laboratory and under real weather conditions. The samples were exposed in the soil for 4, 7, 9 and 11 days. An analysis of mass loss and changes in their mechanical properties, compared to unexposed samples, were determined.

1. Introduction

Agro textiles nonwovens are most often used for laying over crops in open fields and closed areas to accelerate crop growth, protection from frost, low temperatures and insects, and its application reduces the need for herbicides and pesticides and they are often made of synthetic polymers such as polypropylene and polyethylene. Despite the fact that synthetic polymers last century were considered the fibers of the future, today they represent a major environmental problem. A major disadvantage of synthetic polymers is the long-term decomposition process and the problem of removing their waste that accumulates in the environment [1, 2].

Nowadays, there is a growing interest in the use of biodegradable materials, which has resulted in increased the use of natural fibers and bio-based polymers. The term biodegradation is becoming important, as it converts materials into water and minerals, carbon dioxide and new biomass through a biological metabolic process, which presence does not harm the environment. The biodegradation of materials depend upon the polymer chemistry and the environment in which they are exposed. When textiles are buried in soil, soil microorganisms play a major role in the biodegradation process of textile materials, and the biodegradation analysis is often taken as a standard measurement for the environmental friendly textile materials. Some of the important factors that affect the biodegradation time are the presence of microorganisms, the availability of oxygen and water, temperature, humidity and pH of the soil [3-5].

The biodegradability behavior of polymers depends on some physical and chemical properties. The hydrophilic character, crystalline and amorphous structures, the linearity or branching of polymers, molecular compositions and chemical bonds are determinative elements for the biodegradability of polymeric materials. Today, one of the most important biopolymers is PLA mainly due to its high mechanical and thermal properties, very simple production from lactic acid monomer, which is obtained from corn starch [6]. Starch is an attractive raw material for industrial exploitation due to its availability, biodegradability and low price. Degradation of starch (from corn, wheat, rice) and sugar (from sugar beet, molasses) can produce lactic (lactic) acid. Polylactic acid (PLA), is a type of aliphatic polyester that may be made entirely from renewable resources, it is compostable because it dissolves quickly by simple hydrolysis under the right conditions [7-9]. PLA biopolymers has a number of characteristics that are similar to many other thermoplastic fibers, such as controlled crimp, smooth surface and low moisture regain. One unique property in comparison is that it is the only melt-process able fiber from annually renewable natural resources [10, 11]. PLA is ideally suited for many applications in the environment where recovery of the product is not practical, such as agricultural mulch films and bags [12].

Natural cellulose fibers and cellulose regenerates contain cellulose, but differ in chemical composition, crystallinity, degree of polymerization and method of fiber production. Furthermore, natural cellulose fibers, in addition to cellulose, also contain various non-cellulosic admixtures in different percentages. All these differences affect the degradation behavior, which results in different biodegradability. Natural cellulose fibers are known for its wide use in different textile products for different purposes. They are considered environmentally friendly fibers that are known to degrade very easily by soil of microorganisms when they are buried. However, not all cellulose fibers are characterized by the same level of biodegradability due to mutual differences in chemical structure and physical characteristics [6, 13, 14].

Knowledge of the rate and magnitude of biodegradation of natural fibers enables the selection of suitable fibers for the production of biodegradable agro textile nonwovens. In addition, it is also necessary to know the biodegradation in real conditions of use, which mostly depends on climatic conditions that cannot be influenced. The aim of this paper is to investigate biodegradation properties of PLA (as a bio-based polymer), hemp, jute (as natural bast fibers) and viscose (as a cellulose regenerates) for agro textile nonwovens production conducted under controlled conditions in laboratory and under real weather conditions.

2. Material and methods

For the purpose of investigations, an analysis of the biodegradation time of PLA (as a bio-based polymer), hemp, jute (as natural bast fibers) and viscose (as a cellulose regenerates) were performed by soil burial test (*ISO 11721 – Textiles – Determination of resistance of cellulose-containing textiles to microorganisms – Soil burial test*) conducted under controlled conditions in laboratory and under real weather conditions. The samples were exposed in the soil for 4, 7, 9 and 11 days.

Along with determining the mass loss (%), an analysis of changes in their mechanical properties, compared to unexposed samples, were determined on tensile testers Vibroscop and Vibrodyn 400 (Lenzing Instruments, Austria). Used standards and regulations are adapted to testing technical bast fibers. For this purpose, cogged steal clamps are placed on the standard clamps, and the selected testing speed was 3 mm/ min. The selected gauge length of the sample is 5 mm to ensure that all elementary fibers are included in the tested sample during fineness and tenacity testing. Due to the non-homogeneity of bast fibers, the number of measurements was increased (from 50 to 100) and determined according to statistical indications of the degree of reliability. Fineness and tenacity for the samples of Viscose and PLA were determined according to the standards (*HRN EN ISO 1973:2021, HRN EN ISO 5079:2020*). Measurements of fibers of tested properties were performed on conditioned samples.

The protocol of investigation was carried out as it is shown in the Figure 1.

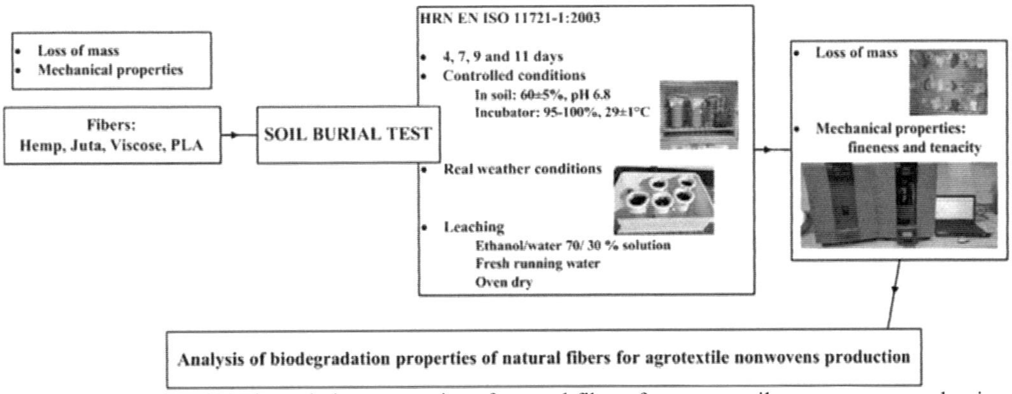

Figure 1. Analysis of biodegradation properties of natural fibers for agro textile nonwovens production

3. Results and discussion

One of the typical analysis performed after burial in soil is the evaluation of the mass loss after burial in the soil and mechanical properties (finesses and tenacity). Mass loss results of samples after burial in the soil are shown in Table 1, and expressed as a percentage.

Table 1. Mass loss results of samples after burial in the soil expressed as a percentage (Δ, %)

Δ, %	4 days		7 days		9 days		11 days	
	L	W	L	W	L	W	L	W
PLA	1.06	0.15	3.31	0.10	2.21	0.43	1.02	0.23
Hemp	36.54	12.08	38.7	17.78	36.23	16.64	52.95	18.04
Jute	13.52	8.01	11.59	10.21	10.52	7.38	13.17	10.04
Viscose	2.80	1.08	3.64	1.78	7.11	3.79	9.41	3.02

The results of mass loss are not in accordance with expectations, that the mass loss of samples increases as the time of exposure to microorganisms increases. It is assumed that illogicality in the obtained results of samples mass loss after exposure to microorganisms is a consequence of the residual soil in the fiber samples so proper conclusions cannot be made. But regardless of the obtained results, it can be observed that a greater reduction in mass after exposure to microorganisms occurred in samples that were exposed to microorganisms under controlled conditions where all desirable conditions for the activity of microorganisms were enabled.

Table 2 show the mean values of the investigated fibers fineness with coefficient of variation, %, conducted under controlled conditions in laboratory and under real weather conditions. Since individual measurements are concerned, a relatively high variation coefficient of bast fibers (ranging from 28.44 % to 44.03 %) was found. The reason can be found in the great non-homogeneity of bast fiber morphology. Test results show a slight change in the fibers fineness, but without the possibility of making a conclusion. It can be assumed that the expected decrease in fineness, after burial in the soil and exposure to microorganisms, does not occur because the biodegradation process occurs in the form of a decreasing the degree of polymerization, i.e. degradation of polymer molecules and not in the form of fiber splitting.

The analysis of the measurement results of samples tenacity after burial in the soil and exposure to microorganisms are shown in Table 3, with coefficient of variation (CV, %) and tenacity loss (Δ, %) compared to the control samples.

Table 2. Fiber fineness results before and after burial in the soil with coefficient of variation

Fineness	Control samples	4 days		7 days		9 days		11 days	
		L	W	L	W	L	W	L	W
PLA									
x̄, dtex	6.84	6.97	7.09	6.62	6.71	7.28	7.17	6.75	6.66
CV, %	9.57	8.47	8.96	8.82	13.92	9.80	12.36	8.61	8.66
Hemp									
x̄, dtex	58.54	60.26	60.24	58.55	60.18	59.01	66.91	65.93	61.78
CV, %	20.58	29.97	37.16	29.28	28.44	31.95	28.67	31.29	30.58
Jute									
x̄, dtex	31.02	32.68	33.06	28.49	29.50	30.04	30.93	32.12	28.65
CV, %	39.19	50.77	41.92	44.03	39.25	38.04	39.27	37.22	42.51
Viscose									
x̄, dtex	1.78	1.70	1.80	1.83	1.80	1.87	1.78	1.80	1.88
CV, %	8.99	13.30	12.72	13.57	9.94	13.23	14.54	12.05	9.85

** Arithmetic mean, $\bar{x}$; coefficient of variation, CV; controlled conditions in laboratory, L; under real weather conditions, W*

The analysis of the measurement results of tenacity (Table 3) after exposure of samples to microorganisms shows an expected reduction of the tenacity of buried samples. Greater reductions in tenacity was obtained in samples exposed to microorganisms under controlled conditions where the better favorable conditions for the activation of microorganisms are ensured such as the availability of oxygen and water, temperature, humidity and pH of the soil. While exposing the samples to microorganisms under real conditions the decrease in tenacity is less because conditions of exposing to microorganisms depends on current climatic conditions.

Table 3. Fiber tenacity results before and after burial in the soil with coefficient of variation and tenacity loss expressed as a percentage (Δ, %)

Tenacity	Control samples	4 days		7 days		9 days		11 days	
		L	W	L	W	L	W	L	W
PLA									
x̄, cN/tex	17.86	13.90	14.67	13.80	13.28	12.83	13.27	12.25	12.73
CV, %	30.04	17.59	13.99	23.92	19.76	27.84	17.94	28.89	20.91
Δ, %	/	-22.17	-17.86	-22.73	-25.64	-28.16	-25.69	-31.41	-28.72
Hemp									
x̄, cN/tex	62.47	42.03	49.45	39.63	45.11	33.36	43.33	31.38	37.17
CV, %	35.18	35.51	37.79	46.34	36.82	46.57	38.24	39.09	40.82
Δ, %	/	-32.72	-20.84	-36.56	-27.79	-46.59	-30.64	-49.77	-40.50
Jute									
x̄, cN/tex	43.94	39.35	39.83	37.97	39.68	34.44	39.54	30.91	39.26
CV, %	28.30	35.27	41.09	34.23	33.61	40.85	38.46	36.72	35.80
Δ, %	/	-10.45	-9.35	-13.59	-9.69	-21.62	-10.01	-29.65	-10.65
Viscose									
x̄, cN/tex	21.36	19.23	20.69	18.55	20.51	17.86	20.25	17.57	17.75
CV, %	10.73	21.54	18.91	22.20	23.25	20.58	22.55	23.20	22.01
Δ, %	/	-9.97	-3.14	-13.15	-3.98	-16.38	-5.19	-17.74	-16.90

** Arithmetic mean, $\bar{x}$; coefficient of variation, CV; tenacity lost, Δ; controlled conditions in laboratory, L; under real weather conditions, W*

PLA degradation upon disposal in the environment is challenge because PLA is largely resistant to attack by microorganisms in soil. But under conditions of high temperature and high humidity, PLA will degrade quickly and disintegrate within weeks to months. In this research PLA is in the form of fibres that are free and without protection of the structural parameters characteristic of woven or non-woven materials. Also, favorable conditions for the activation of microorganisms are ensured. Therefore PLA fibres show an unexpectedly large reduction in tenacity. After 11 days exposure to microorganisms, tenacity loss is 31.41 % for sample exposed to microorganisms under controlled conditions and 28.72 % for samples exposed to microorganisms under real weather conditions.

Cellulose is the basic building material of the Hemp and Jute fibers. In addition to cellulose, fibers contain a small proportion of hemicellulose and so-called non-cellulosic components: lignin, pectin, fats and waxes, water, pigments, minerals and ashes [15]. In addition, each cellulose fibers has different non-cellulose content and composition. All these differences influence the degradation behavior of cellulose fibers, resulting in different biodegradability. Hemp fibers showed the greatest biodegradability what could be explained by the fact that Hemp fibers contain the largest amount of cellulose. The obtained measurement results of hemp fibers show tenacity loss of from 32.72% to 49.77% for samples exposed to microorganisms under controlled conditions, and from 20.84% to 40.50% for samples exposed to microorganisms under real weather conditions. Compared to Hemp fibres, Jute fibres shows almost 50% less tenacity reduction after exposure to microorganisms. An expected reduction of the tenacity can be explained with fact that Jute fibers has 55–65% cellulose, which is responsible for the fibers remarkable mechanical properties and around 12% lignin. Hemp fibers, consisting of about 70% of cellulose and 5% of lignin and wax [14, 15].

In contrast to natural cellulose fibres, Viscose consists of 100% cellulose without any other non-cellulosic components. Also, Viscose fibers are hygroscopic and have a high absorption of moisture from the environment. Due to the low degree of polymerization, weaker orientation and lower proportion of crystallinity, in the wet state these fibers swell and characteristically stiffen and deform. Therefore, it is expected the highest biodegradability regardless of its low crystallinity. It can be assumed that the "skin" that forms on the outer part of the fibers during their spinning slows down the process of biodegradation [16]. In order to obtain complete information about the biodegradation of viscose fibers, further research is needed. However, it is evident that there was a decrease in tenacity.

4. Conclusion

In this research, the biodegradability behaviors of PLA (as a bio-based polymer), Hemp, Jute (as natural bast fibers) and Viscose (as a cellulose regenerates) were studied with standard soil burial method for agro textile nonwovens production. For that purpose, the typical analysis were performed, changes in mass loss and mechanical properties.

The analysis of mass loss results shows illogicality in the obtained results of samples mass loss after exposure to microorganisms what can be explained with the residual soil in the fibre samples so proper conclusions cannot be made.

Obtained results of fineness show a slight change but without the possibility of making a conclusion. Considering the fact that biodegradation occurs in the form of degradation of polymer molecules and not in the form of fiber splitting no different results could be expected.

According to the analysis of the tenacity results, it can be concluded that faster biodegradation was obtained at samples that were exposed to microorganisms under controlled conditions, which can be explained by better favorable conditions for the activities of microorganisms. The results obtained from samples that have been exposed to microorganisms under real conditions are also very important considering that agro textile materials are used under real weather conditions.

In addition, these research are expected to provide basic information for the selection of suitable natural biodegradable fibers for the production of agro textile materials. The continuation of the research will enable the investigation of the ecological acceptability of the development of biodegradable non-woven agro textile from natural and renewable sources.

After reviewing the available scientific literature, a small number of studies based on the biodegradability of fibers were observed. Therefore, the knowledge gained from these researches is considered highly valuable in order to fully understand the biodegradability of fibers and materials. In order to get a complete knowledge about biodegradation of natural fibers, additional research is needed, i.e. morphological, thermal and chemical properties of fibers after exposure to microorganisms.

Acknowledgment
This work has been supported by the European Union from the European Regional Development Fund under the project KK.01.2.1.02.0270 Development of biodegradable nonwoven agrotextiles from natural and renewable sources.

References

[1] Gabrys T, Fryczkowska F, Grzybowska-Pietras J and Binias D 2021 *Mater.* **14** 4335
[2] Tan Z, Yi Y, Wang H, Zhou W, Yang Y and Wang C 2016 *Appl. Sci.* **6** 147
[3] Arshad K, Skrifvars M, Vivod V Volmajer Valh J, Vončina B 2014 *Tekstilec* **57** 118
[4] C Zambrano M, J Pawlak J and A Venditti R 2020 *BioResources* **15** 9786
[5] Čolnik M, Hrnčić Knez M, Škerget M, Knez Ž 2020 *Chem. Ind. Chem. Eng. Q.* **26** 401
[6] Sular V and Devrim G 2019 Fibres Text. East. Eur. 27 100
[7] Teixeira S, Eblagon K M, Miranda F, Pereira R, Figueiredo J L 2021 *J. Carbon Res.* **7** 42
[8] Avinc O and Khoddami A 2009 *Fibre Chem.* 41 391
[9] Li G, Zhao M, Xu F, Yang B, Li X, Meng X, Teng L, Su F, Li Y 2020 *Mol.* **25** 5023
[10] Smith R 2005 *Biodegradable Polymers for Industrial Applications* (Woodhead Publishing)
[11] Bastioli C 2005 *Handbook of Biodegradable Polymers* (Rapra Technology Limited)
[12] Farah S, griffith Anderson D and S Langer R 2016 *Adv. Drug Deliv. Rev.* **107** 367
[13] Mohaptra H S and Malik R K 2015 *J. Textile Sci. Eng.* **6** 1
[14] Hee Park C, Kyung Kang Y and Soon S 2004 *J. Appl. Polym. Sci.* **94** 248
[15] Marques G, Rencoret J, Gutierrez A and C del Rion J 2010 *Open Agric.* **4** 93
[16] Čunko R and Andrassy M 2005 *Fibers* (Zrinski, Zagreb, Croatia)

Novel lyocell fibre from Alberta-grown hemp

L M Degenstein*[1], D King[2], L Lawson[3], W Chute[4], P I Dolez[5]

[1, 2, 3] Davey Textile Solutions, Edmonton, AB, Canada
[4] Techfibre Industries, Edmonton, AB, Canada
[5] Department of Human Ecology, University of Alberta, Edmonton, AB, Canada

Email: ldegenstein@daveytextiles.com

Abstract. Lyocell is a man-made, regenerated cellulosic fibre developed through cellulose dissolution in non-derivative solvents. Hemp offers a preferential source of cellulose for lyocell production as it is considered an environmentally friendly agricultural crop, requiring less water, fertilizers, pesticides, and herbicides than other crops and sequesters carbon dioxide (CO_2) from the atmosphere. There are currently no Canadian sources of domestically manufactured lyocell filament or staple fibres. Our goal is to manufacture 100% lyocell fibre and to demonstrate techniques for pulping and solution-spinning hemp-based lyocell fibre. Utilizing both the bast and hurd of the hemp plant allows for the creation of many different fibre densities with varying properties while contributing to whole plant utilization and Canadian industry creation and expansion.

1. Introduction

Within the last 50 years, global fibre production has grown to over 100 million tonnes annually, contributing to greenhouse gas emissions, microfibre release, and freshwater pollution [1,2]. Polyester and cotton currently dominate the fibre market; however, both fibres have negative environmental impacts as polyester is made from petrochemicals and is non-biodegradable at end-of-life while cotton is a water- and chemical-intensive crop grown on limited arable land [1,3]. Regenerated cellulosic fibres such as lyocell, rayon (viscose), and modal offer alternative fibre options that require less water and energy consumption than cotton and polyester fibre production, respectively [1].

Lyocell is a man-made, regenerated cellulosic fibre developed through cellulose dissolution in non-derivative solvents [4,5]. It is advantageous over other regenerated cellulosic fibre processes (e.g., viscose) as the degree of polymerization is conserved and the N-methyl morpholine-N-oxide (NMMO) solvent is non-toxic and greater than 99% recoverable [4,6]. Lyocell fibres are typically made from cellulosic sources such as wood, bamboo, and eucalyptus pulps [4]; however, the use of these materials can lead to deforestation [7,8]. Hemp offers a preferential source of cellulose as it is considered an environmentally friendly agricultural crop, requiring less water, fertilizers, pesticides, and herbicides than other crops and sequesters carbon dioxide (CO_2) from the atmosphere [9].

There are currently no Canadian sources of domestically manufactured lyocell filament or staple fibres; the textile industry relies on imports, which are often untraceable from both an environmental and social perspective. Global demand for lyocell fibre is expected to grow from 415,600 tonne/year in 2018 (~$1B) to 745,900 tonne/year by 2027 (~$2B) [10]. Significant markets for lyocell fibre include consumer textiles, personal protective equipment (PPE), medical, hygiene, automotive and aerospace industries owing to its absorbency (11.5% moisture regain), high wet (4.2-4.6 g/denier) and dry tenacity (4.8-5.0 g/denier), handle, and biodegradability [6,11]. By manufacturing lyocell from hemp, we will be assisting in whole plant utilization thus making growing agricultural hemp for multiple purposes more economically sustainable. Considering this environmental and economic context, our goal is to manufacture 100% lyocell fibre and to demonstrate techniques for pulping and solution-spinning hemp-based lyocell fibre.

2. Materials and Methods

Two parts of the hemp plant can be utilized for fibre production — the bast (bark) and the hurd (shive) [12]. Following the same principles as other regenerated cellulosic fibre production steps [6], bast is removed from hurd via chemical retting and decortication which breaks down lignin and pectin for easier processing. The bast and hurd fibres can move directly to the yarn spinning process as staple fibre or can be dissolved in NMMO solvent. The dope solution is then spun, washed, treated for desirable properties, dried, and extruded as lyocell filament fibre (Figure 1). Using high alpha cellulose pulps from hemp allows us to produce high tenacity lyocell fibres that will be ideal for use in PPE applications (i.e., industrial use clothing). In addition, previous research has demonstrated the ability to impart flame resistant (FR) properties onto solvent-spun cellulosics via phosphorylation treatments [13]. Imparting FR properties as an additive to the dope solution will result in FR fibres for use in industrial protective apparel. Furthermore, the differing viscosities (i.e., degrees of polymerization) of dissolving pulps produced from hemp bast and hurd offer unique opportunities to create fibre structures with improved organoleptic and technical performance from a single plant source, which is not possible from plants such as trees (e.g., eucalyptus) or bamboo.

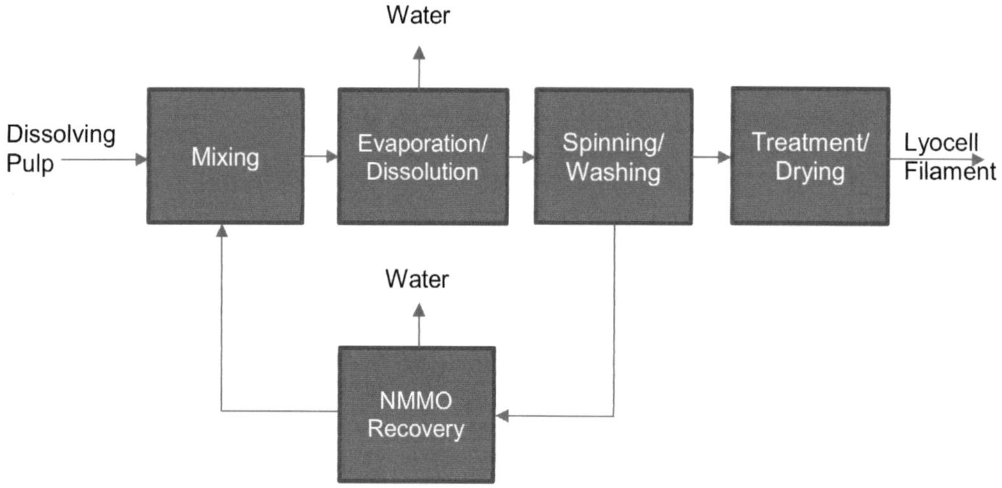

Figure 1. Hemp-based lyocell fibre production process.

Previous work completed at Thuringian Institute of Textile and Plastics Research (TITK) and Saxon Textile Research Institute (STFI) have demonstrated proof of concept of chemically pulping hemp bast fibre, as well as extruding the pulp solution via the lyocell process [14,15]. Building on this work, Davey Textile Solutions determined the feasibility of lyocell from hemp in Canada alongside Techfibre, who provided initial characterization of Alberta-grown hemp dissolving pulps, Econotech, who delivered technical evaluation and testing, and College Technology Transfer Centre at Cégep de Saint-Hyacinthe (CTT Group), who provided an in-depth investigation into the viability of FR additives. With proof of concept and initial testing completed, we want to move forward with our project of a pilot scale lyocell plant. We are aiming to set up a hemp lyocell pilot facility capable of processing approximately 12 tonnes of hemp fibre/year. This pilot facility will prove the efficacy of the process with the goal to scale up to profitable commercialization within 2 years of pilot operation.

3. Results and Discussion

Canadian hemp is a readily accessible commodity farmed for various usages. Hemp lyocell fibre, produced in Alberta, would contribute to a target of whole hemp plant utilization (i.e., using residual whole straw from oilseed production), as well as establish the first Albertan (and Canadian) owned and operated extruded cellulose fibre manufacturing facility [16]. Furthermore, growing agricultural hemp for multiple purposes aids in economical sustainability.

Lyocell fibres derived from residual hemp straw have a built-in sustainability story that offers greater global acceptance than those produced from wood pulps. Growing hemp is capable of sequestering 22 tonnes of CO_2e/acre, and "every tonne of industrial hemp contains 0.445 tonne carbon absorbed from the atmosphere (44.46 % of stem dry weight)" [9] (p. 3). Based on this calculation, if we use 12 tonnes of hemp/year for a pilot project, this equates to 5.34 tonnes of carbon. At commercialization, we would realize higher volumes of CO_2 sequestration. This will be the first available FR fibre produced in Canada for a market that consumes 5,000 tonnes of coveralls per year in Alberta alone, based on our previous investigations.

4. Conclusion

Lyocell from hemp provides a novel opportunity for fibre production in the face of our global sustainability challenges. We are developing a process to manufacture hemp bast and hurd fibre from residual oilseed straw into novel lyocell filaments. As textile products are one of the leading contributors to greenhouse gases, we will be changing the direction of this industry by manufacturing lyocell from carbon-sequestering hemp and contributing to Canadian industry creation and expansion.

Acknowledgements

The authors would like to express appreciation for the support from: Davey Textile Solutions, Techfibre Industries, University of Alberta, CTT Group, TITK, STFI, Econotech, and InnoTech Alberta.

References

[1] Niinimäki K, Peters G, Dahlbo H, Perry P, Rissanen T and Gwilt A 2020 The environmental price of fast fashion *Nat. Rev. Earth Environ.* **1** 189–200 (https://doi.org/10.1038/s43017-020-0039-9)

[2] Kissinger M, Sussmann C, Moore J and Rees W E 2013 Accounting for greenhouse gas emissions of materials at the urban scale – relating existing process life cycle assessment studies to urban material and waste composition *Low Carbon Econ.* **4**(1) 36–44 (http://dx.doi.org/10.4236/lce.2013.41004)

[3] Bevilacqua M, Ciarapica F E, Mazzuto G and Paciarotti C 2014 Environmental analysis of a cotton yarn supply chain *J. Clean. Prod.* **82** 154–165 (https://doi.org/10.1016/j.jclepro.2014.06.082)

[4] Jiang X, Bai Y, Chen X and Liu W 2020 A review on raw materials, commercial production and properties of lyocell fiber *J. Bioresour. Bioprod.* **5**(1) 16–25 (https://doi.org/10.1016/j.jobab.2020.03.002)

[5] Sayyed A J, Deshmukh N A and Pinjari D V 2019 A critical review of manufacturing processes used in regenerated cellulosic fibres: viscose, cellulose acetate, cuprammonium, LiCl/DMAc, ionic liquids, and NMMO based lyocell *Cellulose* **25**(5) 2913–2940 (https://doi.org/10.1007/s10570-019-02318-y)

[6] White P 2001 Lyocell: the production process and market development (*Regenerated Cellulose Fibres*) ed C Woodings (Cambridge, UK: Woodhead Publishing) pp 62–87

[7] Briggs J. Can hemp replace trees as a major source for paper? [Internet]. Hemp Frontiers; 2005 [updated 2005 May 23; cited 2022 Apr 28]. Available from https://hempfrontiers.com/hemp-paper-vs-tree-paper/

[8] Chua, J M. Amazon pledges to rid private-label brands of forest-harming viscose [Internet]. Sourcing Journal; 2020 [updated 2020 Jul 7; cited 2022 Apr 28]. Available from https://sourcingjournal.com/topics/raw-materials/amazon-private-label-canopystyle-deforestation-sustainable-viscose-2022-219512/

[9] Vosper J. The role of industrial hemp in carbon farming [Internet]. New South Wales: GoodEarth Resources; 2011 [cited 2022 Apr 28]. 6 p. Available from https://www.aph.gov.au/DocumentStore.ashx?id=ae6e9b56-1d34-4ed3-9851-2b3bf0b6eb4f

[10] Business Wire. Global lyocell fabric market to reach 745.9 KT by 2027 – coherent market insights. Seattle WA: Business Wire; 2019 [updated 2019 Nov 25; cited 2022 Apr 28]. Available from https://www.businesswire.com/news/home/20191125005611/en/Global-Lyocell-Fabric-Market-Reach-745.9-KT

[11] Kadolph S 2011 *Textiles* (Upper Saddle River, NJ: Prentice Hall)

[12] Zhao X, Wei X, Guo Y, Qiu C, Long S, Wang Y *et al* 2021 Industrial hemp – an old but versatile bast fiber crop *J. Nat. Fibers* (https://doi.org/10.1080/15440478.2021.1907834)

[13] Belosinschi D, Brouillette F, Shi Y, Paradis J and Douvet J, inventors. Bereskin & Parr LLP/SENCRL, SRL, assignee. *Phosphorylated lignocellulosic fibers, uses and processes of preparation thereof.* Canadian Patent PCT/CA2017/050717, 2017. Available from: https://patents.google.com/patent/WO2017214719A1/en

[14] Kosan B, Meister F, Sigmund I and Paulitz, J 2019 Innovative dissolving pulps for application in cellulose MMF production *Lenzinger Berichte* **95** 9–14

[15] Meister F, Sigmund I, Paulitz J, Kosan B; Kreig M and Mooz M 2017 *56th Dornbirn Man-made Fibers Congress*

[16] Breland P, Graham, L-J. Industrial hemp overview and market opportunities [Internet]., Vermilion/Vegreville: Government of Alberta; 2018 [cited 2022 Apr 28]. 26 p. Available from: https://www1.agric.gov.ab.ca/$Department/deptdocs.nsf/all/bt16408/$FILE/1GrowingIndustrialHempFeb6 VermillionFeb8VegrevilleFinal.pdf

IOP Conf. Series: Materials Science and Engineering 1266 (2023) 012020 doi:10.1088/1757-899X/1266/1/012020

Multifilaments based on partially miscible polymers blend filled with carbon nanotubes

J Regnier[1], C Cloarec[1], A Cayla*[1], C Campagne[1] and E Devaux[1]

[1]Univ. Lille, ENSAIT, ULR 2461 - GEMTEX - Génie et Matériaux Textiles, F-59000 Lille, France
*Corresponding author: aurelie.cayla@ensait.fr

Abstract. Many textile fields, such as industrial structures or clothing, use the electrical conductivity variation of yarns to detect fluid leakage. Such yarns can be developed by melt spinning conductive polymer composites (CPC). CPC filaments are composed of a polymer's matrix which is blended with sufficient quantity of electrically conductive fillers to make the filament conductive. To combine properties or improve the compounds preparation, more and more studies are investigating different polymers blends. In this study, CPC monofilaments and multifilaments are developed and characterized to observe the formulation influence on spinnability and the implementation process on the water detection. Two principles of water detection are studied on the CPC which is composed of a blend of partially miscible polymers (polyethylene terephthalate (PET)/polybutylene terephthalate (PBT)) filled with carbon nanotubes (CNT). The principle of absorption is based on the electrical conductivity variation of the filament in contact with water. For the short circuit principle, the presence of the liquid is detected when the water creates a conductive path between two filaments in parallel.

1. Introduction

Many applications are based on smart textiles via the electrical conductivity variation [1]. For instance, in the environmental or clothing field, it is possible to detect a liquid leakage by using a conductive yarn [2]. This one can be elaborated from Conductive Polymer Composites (CPC) and more specifically from nanocomposites by the melt spinning process. Nanocomposites are insulating polymers blended with electrically conductive nanofillers. Many studies focus on the optimization of these sensors according to their formulation, the process and the solvent [3].

This study deals with a partially miscible polymers blend filled with carbon nanotubes (CNT). The two polymers are the polyethylene terephthalate (PET) and polybutylene terephthalate (PBT). This formulation is developed in different proportions in a form of monofilament and multifilament. The CNT and PBT impact on the implementation process, the electrical and detection properties is investigated.

2. Materials and methods

2.1. Conductive filaments.

Polyethylene terephthalate (PET) is a semi-crystalline thermoplastic polymer, referenced by POLYESTER CHIP 5140 and supplied by INVISTA Resign & Fibers (Hattersheim am Main, Germany). Its density is 1.38 and its melting temperature is 245°C.

IOP Conf. Series: Materials Science and Engineering 1266 (2023) 012020 doi:10.1088/1757-899X/1266/1/012020

Polybutylene terephthalate (PBT) is also supplied by INVISTA Resign & Fibers (Hattersheim am Main, Germany) and is referenced by PBT 2000. PBT is like PET, that means a semi crystalline thermoplastic polymer of the polyester family. It has a density of 1.34 g/cm3, a melting temperature of about 230°C and a melt index of 0.8 dl/g.

Multiwalled carbon nanotubes (MWCNT) are used to provide the conductive property to each blend. MWCNTs are produced by the company Nanocyl (Sambreville, Belgium) and referenced by NC 7000. They have an average length of about 1.5 µm, a diameter of 9.5 nm, a specific surface area of 250 to 300 m^2/g and a volume electrical resistivity of about 10-4 Ω/cm. The notation for the paper will be only CNT.

2.2. Compounds preparation.

2.2.1. Extrusion. The compounds preparation is carried out by a first extrusion to fill the PET with the CNT in order to disperse the fillers well. The CNT proportion varies from 0 to 5 wt.% in the polymer. To simplify the notation, the CNT content (X wt.%) in the PET will be noted: PET_{XCNT}. All the nanocomposites are implemented by a co-rotating intermeshing twin-screw extruder from Process 11 Parallel Twin-Screw Extruder from Thermofischer (Waltham, Massachusetts, USA) with a barrel length of 40:1 L/D. The rotating speed of this extruder is 100 RPM, and the temperatures profile is: 215 °C – 275 °C – 280 °C – 280 °C – 275 °C – 270 °C – 265 °C – 265 °C.

A second extrusion step is then processed in order to incorporate the PBT inside the nanocomposite in different proportions (from 0 to 40 wt.%). The formulation will be noted $PET_{XCNT}Y_1/PBTY_2$ (with Y_1 the wt.% PET_{XCNT} and Y_2 the wt.% PBT). The same process parameters and temperatures profile are applied for the second extrusion. Before each extrusion, the granulated polymers are dried at 80 °C for 16 h. The monofilaments have a diameter of approximately 1.2 ± 0.3 mm.

2.2.2. Melt spinning process. Some PET_{CNT}/PBT formulations are developed as multifilament by using the Spinboy I spinning pilot of Busschaert Engineering (Outrijve, Belgium). The pellets of the previously extruded blends are dried for 5 h at 140 °C to remove moisture. The single screw of the pilot has an L/D ratio of 30. The temperatures profile is: 260 °C - 262 °C - 264 °C - 265 °C - 265 °C - 265 °C. At the end of the single screw, the molten blends pass through two dies, each with 40 holes of 1.2 mm diameter. The two bundles of monofilaments are then brought together by sizing and stretched by a heating roller system.

2.3. Methods of characterization.

To investigate the CNT and PBT influence on the water detection properties, two detection principles are established: the absorption principle and the short circuit. They are both based on the electrical conductivity variation of a filament in contact with water. Ten tests that are conditioned at 20 °C and a humidity of 35 % for each yarns and principles, are necessary to validate the results and the reproduction of these protocols.

2.3.1. Morphology. The Scanning Electron Microscope (SEM) images permits to observe the blends morphology. First, the samples are prepared by cryofracture: frozen under nitrogen and then broken. They are observed by an Environnemental Quanta 200 FEG Microscope (FEI Compagny) at 5 kV and different magnifications.

2.3.2. Rheological properties. The Melt Flow Index (MFI) determines the fluidity of a polymer and more precisely the flowing polymer weight in 10 min at the implementation temperature. According to the standard ISO-11333, the test is carried out on the melt flow tester from ThermoHaake with a temperature of 265 °C and a pressure of 2.16 kg. Before the test, the polymer pellets are dried at 80 °C for 16 h.

2.3.3. Electrical and detection properties. The monofilaments are connected to a Keithley 2461 SourceMeter (Beaverton, OR, USA) to quantify the electrical conductivity of the different blends. The current intensity is measured for an applied voltage ranging from -0.5 V to 15 V with an increment of 0.1 V. The data of the current as a function of the voltage sent permits to determine the electrical conductivity by the equation 1:

$$\sigma = \frac{L}{RS} \tag{1}$$

Where σ is the electrical conductivity of the filament (S/m), L is the filament's length (L = 0.1 m), S is the filament's area (m^2) and R is the resistance measured (Ω).

The detection principle by absorption is based on the electrical conductivity variation of the filament after 24 h in contact with water, as shown in figure 1(a). The conductivities of the filament before and after immersion are calculated. To observe the influence of mass fractions, the rate of change of electrical conductivity between the dry and the immersed filament is called the sensitivity and noted Sw. It is calculated by the equation 2:

$$Sw = \frac{\sigma i - \sigma f}{\sigma i} \times 100 \tag{2}$$

Where Sw is the water detector sensitivity (%), σi is the initial electrical conductivity (of the dry filament) (S/m) and σf the electrical conductivity of the immerged filament (S/m).

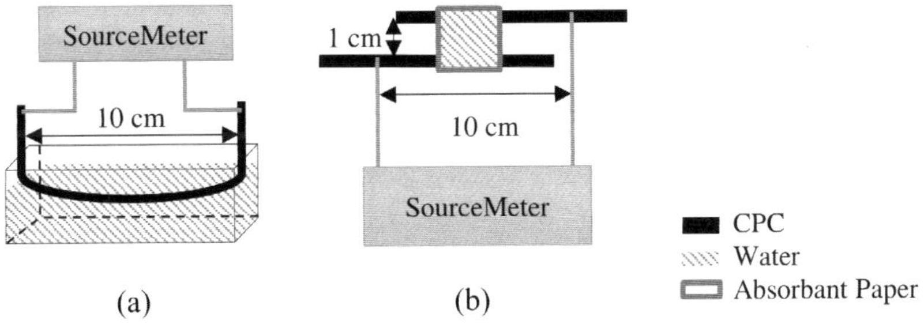

Figure 1. Principle of absorption (a) and the short circuit (b).

The short circuit uses a drop of water to create a conductive path between two parallel filaments as shown in figure 1(b). The water closes the circuit between the two conductive filaments and triggers the electrical signal [4]. To observe the formulation influence, the conductance of the circuit is calculated with the equation 3. To make the test repeatable, the water is deposited on a square of 3 × 3 cm of absorbent paper which links the parallel filaments. It permits to overcome the shape of the drop and the problems of absorption that could vary the conductance of the circuit.

$$G = \frac{1}{R} \tag{3}$$

Where G is the conductance of the circuit (S) and R the resistance of the circuit (Ω).

3. Results and discussion

3.1. Morphology and rheological property

The first step of characterization is the morphology. It is observed by SEM images. The SEM image of $PET_{1CNT}90/PBT10$ in figure 2 suggests a good dispersion of nanotubes in the blend. Moreover, it is possible to expect a partial miscibility of the PBT and PET, which can be confirmed by DSC analysis.

Figure 2. SEM images of the $PET_{1CNT}90/PBT10$ blend (5 µm scale).

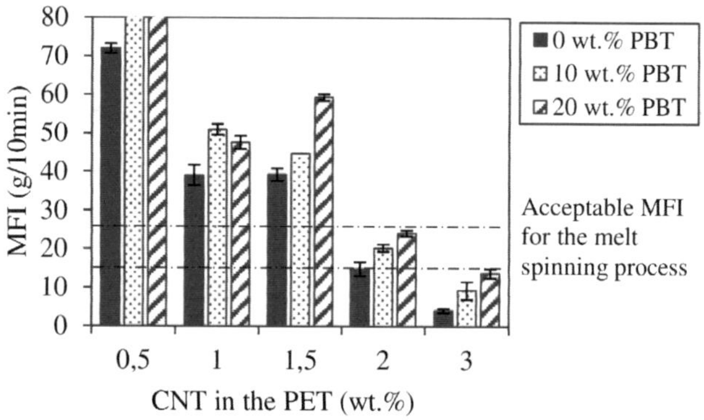

Figure 3. Melt Flow Index (MFI) at 265 °C and 2.16 kg of the blends with different proportions by weight of PBT and CNT.

In order to be able to spin the nanocomposite, its fluidity is then studied through the Melt Flow Index at a fixed pressure and temperature. In Figure 3, the introduction of CNT into PET significantly decreases the fluidity of the polymer, from 72 to 4 g/10 min for 0 to 3 wt.% CNT. Above 4 wt.% CNT, the measurement could not be performed at this temperature due to a significant loss of fluidity. By hypothesis, the introduction of CNT in the PET and the formation of the conductive network reduce the mobility of the polymer macromolecular chains.

However, PBT seems to increase the fluidity of the nanocomposite by limiting the impact of nanotubes [5]. Therefore, it necessary to add PBT in the nanocomposite to improve the elongational viscosity for the melt spinning process. For instance, the addition of 10 wt.% in the PET_{2CNT} allows to obtain an acceptable MFI value for melt spinning at 265 °C, i.e. 20.3 g/10min. For PET_{3CNT}, the incorporation of PBT does not permit to reach the acceptable limits for melt spinning. Therefore, the processing temperature must be modulated by increasing it by 18 °C to spin the blends.

3.2. Electrical property

Figure 4 highlights the electrical percolation threshold of the PET_{CNT}/PBT blend. It shows the electrical conductivity of the monofilaments as a function of the mass fractions of fillers and PBT in the polymer. For all PBT proportions, it is possible to observe a leap in conductivity between 0.8 and 1.5 wt.% CNT. This strong variation indicates the percolation threshold. It means when the blends switch from an insulating to electrically conductive. The best balance must be found between the addition of PBT for spinnability and the CNT to ensure the electrical property after the spinning process. This compromise is 2 or 3 wt.% CNT and a minimum of 10 wt.% PBT.

IOP Conf. Series: Materials Science and Engineering 1266 (2023) 012020 doi:10.1088/1757-899X/1266/1/012020

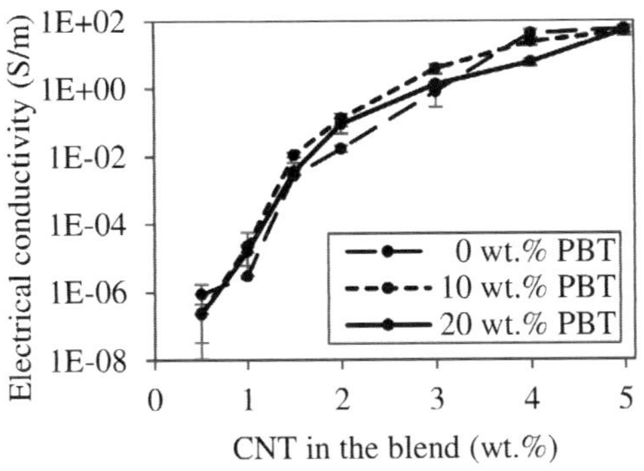

Figure 4. Initial electrical conductivity of the monofilaments.

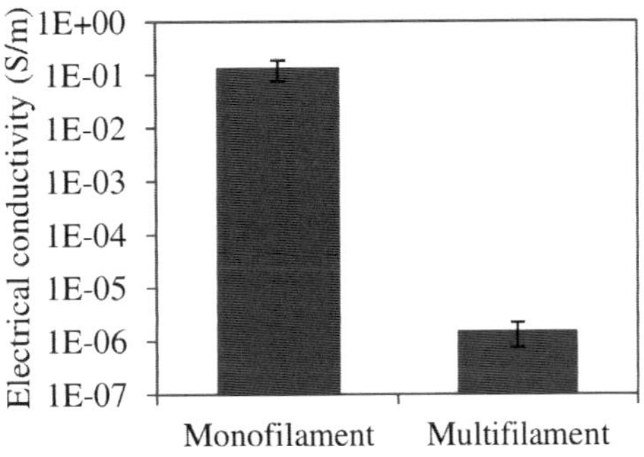

Figure 5. Initial electrical conductivity of the $PET_{2CNT}90/PBT10$ according to the implementation process

However, in Figure 5, the electrical conductivity of the $PET_{2CNT}90/PBT10$ multifilament drops sharply compared to the monofilament. The results may be explained by the stretching during the melt spinning process. The distance between the charges increases and may not allow the formation of conductive networks.

3.3. Detection properties

The monofilaments and multifilaments are used for the water detection. These detections are tested with demineralized water at room temperature. The ionic conductivity is about 2.3 ± 0.2 µS/cm and its surface tension of 71.3 ± 1 mN/m. Theses parameters are controlled before each test to reduce water-related uncertainties and ensure the protocols repeatability.

3.3.1. Short circuit principle

The short circuit signal depends on several parameters: the distance between the two parallel filaments and the properties of the water, which are both fixed, as well as the properties of the filament. The electrolysis of water allows the transport of H+ ions between the anode and cathode. Therefore, in this study, the drop of water is only seen as a switch that closes the electrical circuit by its contact with the two filaments. The figure 6 shows the short circuit conductance according to the CNT and PBT ratio.

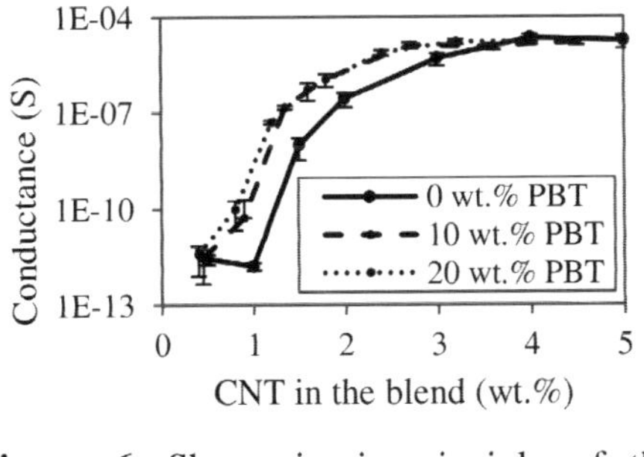

Figure 6. Short circuit principle of the monofilaments

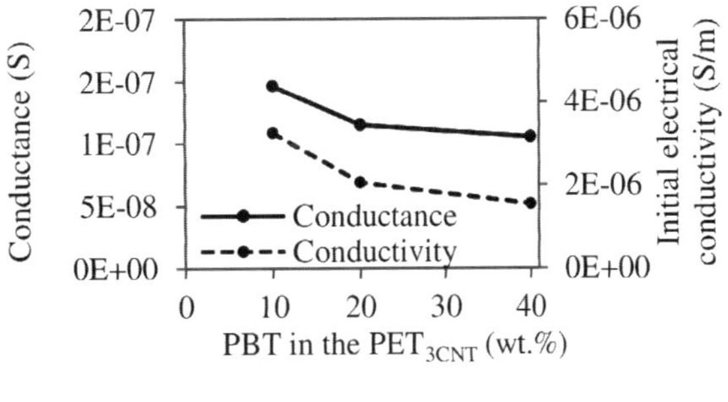

Figure 7. Short circuit principle of the multifilaments

The detection signal is about 10^{-7} to 10^{-5} S for the blends which containing more than 2 wt.% CNT. As with the initial conductivity, the conductance increases sharply at the percolation threshold. For blends with low initial electrical conductivity, the conductance is about 10^{-13} S. Therefore, the signal seems to be only correlated with the initial conductivity of the monofilament. This hypothesis can be also observed on the figure 7. According to the PBT proportion, the short circuit conductance of the multifilaments seems to have approximatively the same variation as their initial electrical conductivity.

3.3.2. Absorption principle

The influence of the formulation and the implementation process is observed on the water detection. The figure 8 shows the sensitivity as a function of the filler content and initial electrical conductivity. It highlights the increase in sensitivity of formulations which are near the percolation threshold. The high concentration of CNT does not allow an important sensitivity due to the high electrical property of the monofilaments. At lower concentration, around the percolation threshold, the sensitivity is more important. The high standard deviations represent the instability of the initial conductivity values that can be observed for blends around the percolation threshold.

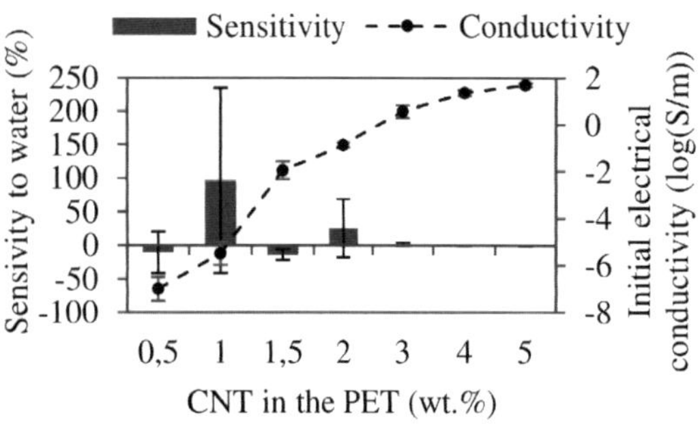

Figure 8. Sensitivity to water of the monofilaments according to the filler's ratios.

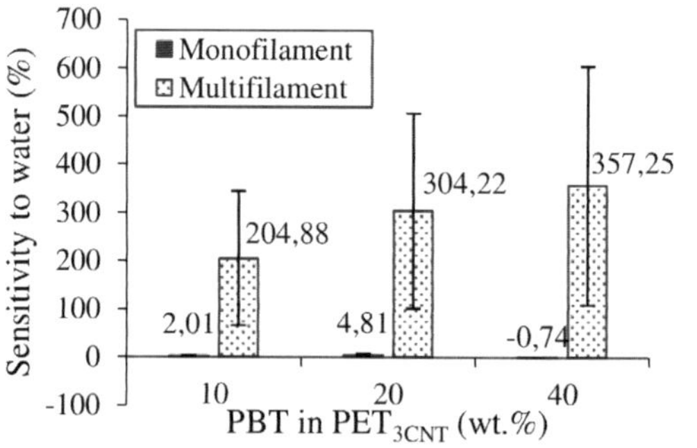

Figure 9. Sensitivity to water of the PET3CNT according to the PBT ratios and the implementation process.

Regarding the figure 9, the form of multifilament increases the sensitivity to water. By hypothesis, the stretching during the melt spinning process and the form of multifilament permit to increase its capillarity property and the contact area between water and the monofilaments of the multifilament.

4. Conclusion

The influence of different formulations on the monofilament's properties is first studied in order to select the most optimal formulation for the melt spinning process. Blends which contain 3 wt.% CNT, offer a suitable fluidity for the spinning process and a good signal for the short-circuit. However, monofilaments which are just below the percolation threshold (with 1 wt.% CNT), have the best sensitivity to water. A compromise between good fluidity, initial conductivity and water detection has led to develop some multifilaments. After spinning, the initial conductivity of the multifilaments drops sharply in contrast to their water sensitivity.

In conclusion, the sensitivity to water of the multifilament can be refined by modifying the formulation and implementation parameters. The next step is to increase the filler content and thus of PBT in order to ensure the short circuit signal and the spinnability.

References

[1] Castro M, Kumar B, Feller JF, *et al.* 2011 *Sensors and Actuators B: Chemical* **159** 213–219
[2] Goldfeld Y, Rabinovitch O, Fishbain B, *et al.* 2016 *Journal of Intelligent Material Systems and Structures* **27** 469–489
[3] Feller J-F, Petitjean E. 2003 *Macromol Symp* **203** 309–316
[4] Xu S, Ma Q, Yang X-F, *et al.* 2017 *RSC Advances (RSC Publishing)* **7** 41117–41126
[5] Mishra SP, Deopura BL. 1984 *Rheol Acta* **23** 189–194

High-Speed Imaging Analysis of Laminated Fabric Tearing Behaviour

Laura Munevar-Ortiz*[1,2], John A. Nychka[2] and Patricia I. Dolez [1]
[1] Department of Human Ecology, University of Alberta, Canada
[2] Department of Chemical and Materials Engineering, University of Alberta, Canada
E-mail: munevaro@ualberta.ca, jnychka@ualberta.ca, pdolez@ualberta.ca

Abstract. Laminated fabrics are widely used in diverse industries because of their ability to adapt to many different functions. Understanding their tear force behaviour is critical to evaluate their serviceability. The tearing behaviour of laminated fabrics can be examined using high-speed imaging and analyzed using image analysis techniques. This technique is an attractive tool to elucidate the tearing behaviour of laminated fabrics. Through high-speed imaging it is possible to observe the tearing process on the fabric (e.g., yarns bearing the load before failure, yarn breaking behaviour) and membrane (e.g., pop-in crack length, crack propagation, formation of wrinkles, thinning) sides of the laminated fabrics to understand the tearing behaviour. In addition, this technique helps quantify different phenomena such as the membrane pop-in crack length, and crack wake bridging width by yarns. High-speed imaging analysis is a promising technique to visualize the tear phenomena in laminated fabrics.

1. Introduction

Laminated fabrics have attracted widespread attention because of their versatility, which broadens the uses of textiles [1,2]. Among other characteristics, they allow combining liquid resistance and water vapour transmission. Laminated fabrics are a multilayer structure, which usually consists of at least three essential components: a membrane attached to a base fabric with an adhesive [3]. There is wide applicability for laminated fabrics across a number of industries such as agriculture, medical, transportation, construction, geotechnical engineering, home furnishings, packaging, clothing, and sports [1]. An important application of laminated fabrics is in firefighters' protective clothing due to their ability to keep the user dry while allowing perspiration to exit [4,5]. Firefighters' protective clothing in North America usually includes three layers, the middle layer being a laminated fabric called the moisture barrier [6].

Tear strength is one of the most important characteristics of fabrics' mechanical performance; in typical service situations, a fabric will generally fail by tearing [7-9]. A thorough understanding of the tearing behaviour and damage morphology is critical to allow the proper design of laminated fabrics for industrial applications [10]. The tearing behavior of fabrics by the trapezoid procedure has been the subject of several research works [11-16]. According to the results of these investigations, the tearing strength of a fabric by the trapezoid method depends on the geometry of specimens, weaving density, initial distance between the clamps, trapezoid angle, and breaking strength of the yarns. However, these works did not study laminated fabrics, which because of their multi-component nature, exhibit complex physical behaviours due to the inherent component properties and inter-component interactions and constraints.

Failure analysis is a process of making observation, collecting analytical data, and gathering facts to determine how a component failed [17], and is of great importance to identify why the failure of a

IOP Conf. Series: Materials Science and Engineering 1266 (2023) 012021 doi:10.1088/1757-899X/1266/1/012021

material occurred to improve the design, materials, fabrication, inspection methods, and prevent further failures. Some methods of failure analyses include obtaining the material history and information, performing macroscopic and microscopic examination, analysing the material's composition, molecular structure, and physical properties, and reviewing the data [18]. For laminated fabrics, more specifically for moisture barriers, no in-depth failure analysis of the tearing behaviour has been found in the literature.

This paper presents a high-speed imaging technique to macroscopically examine and analyse the tearing behaviour of laminated fabrics as tested during the trapezoid procedure.

2. Materials and Methods

Three moisture barriers commonly used in firefighters' protective clothing were selected for the study (see Table 1).

Table 1. Characteristics of the laminated fabrics studied

Code	Base fabric		Membrane
	Structure	Fibre content	Material*
MB1	Rip stop	93% meta-aramid, 5% para-aramid and 2% antistatic	Microporous ePTFE/FR PU
MB2	Rip stop	93% meta-aramid, 5% para-aramid and 2% antistatic	Bicomponent ePTFE/FR PU coated with FR PU
MB3	Nonwoven	85% meta-aramid and 15% para-aramid	Microporous ePTFE/FR PU

* ePFTE: Expanded polytetrafluoroethylene; FR: Fire-resistant; PU: Polyurethane

The tear force of the moisture barriers was measured according to the standard test method ASTM D5587-15 (2019) [19]. The tear force was determined using a constant-rate-of-extension (CRE) type tensile testing machine (Instron Model 5565, Illinois Tool Works Inc., Norwood, MA) equipped with a 5 kN load cell and controlled by a computer software program (Bluehill; Instron Corp). The rate of crosshead displacement was set to 300 mm/min. Due to limitations in fabric availability, a slightly smaller size (55 x 110 mm) was used for the specimens (Figure 1) which has been previously shown to have no affect on the tear strength measurement results [16]. According to the standard test method, the specimens were conditioned in standard atmospheric conditions (20 ± 2°C; 65 ± 5% RH) for 24h prior to testing. The tear strength test was also carried out at the same conditions.

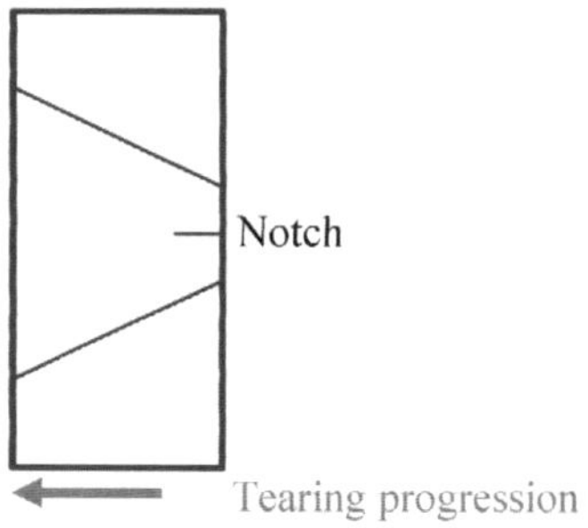

Figure 1. Specimen configuration. Diagonal lines indicate locations of clamping. Tearing initiates at the notch root and progresses from right to left (as indicated by the red arrow).

The tearing process of the laminated fabrics was recorded with a high-speed camera (FastCAM-SA5, Photron, CA) at a frame rate of 1,000 frames per second using two AOS technologies AG A-LED-

W150 light sources to illuminate the entire width of the specimen in the field of view, as shown in Figure 2. A higher magnification (and higher frame rate of 20,000 frames per second) was also used to capture macroscopic details of the tearing zone on each of the fabric and membrane sides. The camera was intermittently moved laterally to follow the tearing zone for each test on each surface, stopping and restarting the tensile tester in between camera movements. Simultaneous recording of the tearing behaviour of both surfaces of the moisture barriers (fabric and membrane surfaces) could not be performed due to space limitations behind the tensile test machine: the recordings from different surfaces and magnified tear zones were performed on separate specimens wherein the examined surface was positioned in the testing machine to face the camera.

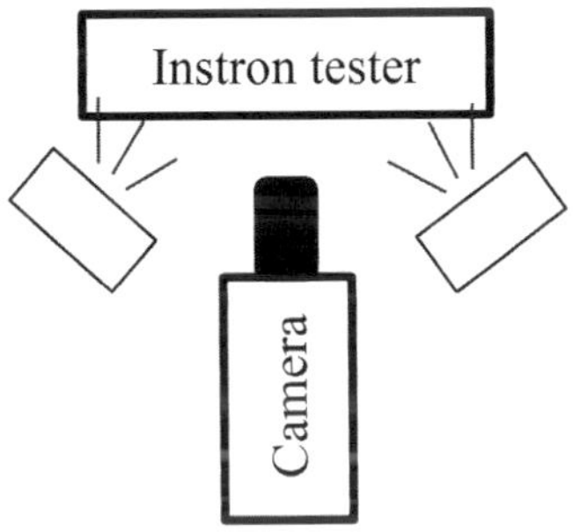

Figure 2. Experimental set-up

The high-speed videos were analyzed via the FASTCAM Viewer Software version 4 (Photron Limited, Japan) software to visualize the different steps of the tearing propagation throughout the moisture barriers, and key frames were chosen and extracted for further image analysis.

3. Results and discussion

High-speed imaging allows detailed visualization of the tearing behaviour by acquiring a series of image frames of the laminated fabrics during tearing. The high-speed recordings aided visualization of different, yet imperceptible behaviours in real-time, on the fabric and membrane sides of the laminated fabrics. For instance, on the fabric side of MB1 and MB2 (Figure 3.a), multiple yarns behind the notch tip share the load before they all break either simultaneously or singly as the tearing progresses from right to left. In Figure 3.b, individual yarns break before the paired yarns. The ripstop structure of the fabric base is used to slow down or stop the progression of the tear and increase the tear strength by adding paired yarns at regular intervals.

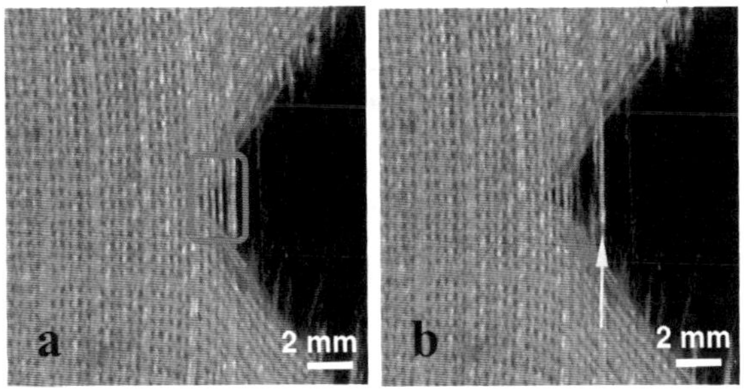

Figure 3. (a) Loaded yarns behind the notch prior to breaking (within red box); (b) Yarn breakage of individual yarns (red arrows) before paired yarns (yellow arrow).

On the membrane side, high-speed imaging enabled several characteristics to be observed among the different laminated fabrics. Figure 4.a illustrates that the crack/notch propagation in the membrane occurs ahead of the tearing of the base fabric. Also, this image shows a rapid pop-in crack on the membrane first, followed by slower crack growth. Figure 4.b showcases the creation of diagonal wrinkles from the clamps to the tearing zone. The formation of wrinkles might potentially be due to stress relaxation after the material has ruptured and is not under the constraint of the fabric; the wrinkles are the membrane's response to the lateral pulling of the test due to the clamping orientation. Figure 4.c shows thinning on the membrane side of one of the laminated fabrics. Thanks to the high-speed imaging it is possible to observe phenomena that are impossible to see at the normal speed of the tear strength test.

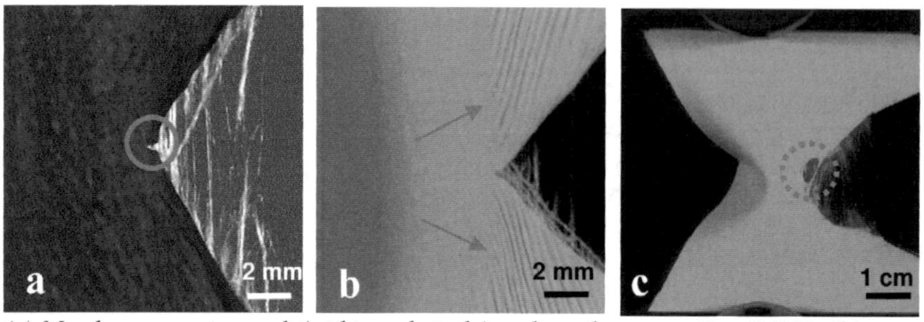

Figure 4. (a) Membrane pop-in crack (within red circle) and crack propagation (MB1); (b) Membrane wrinkles (MB2) indicated by red arrows; (c) Thinning on the membrane side of moisture barrier (MB3) within dashed circle.

To quantify some variables such as the pop-in crack length and the crack wake bridging width by yarns, the FASTCAM Viewer Software version 4 was used. To measure the pop-in crack length, the distances were measured via the function "measurement-two points". This function will give the number of pixels in the selected distance. After measuring the pop-in crack length, the diagonal of the trapezoid (the side where the laminated fabric is attached to the clamps) was measured as the reference distance since it is a known physical dimension. Then, the actual dimension of the pop-in crack was calculated. Figure 5.a shows the pop-in crack distance measurement in the software. The crack wake bridging width by yarns was also measured. Another function that could be used in the software is the "measurement-count" to highlight the number of yarns that the crack wake has, as shown in Figure 5.b. There are other multiple analyses that can be done with the software such as measurement of angles and measurement of load sharing width.

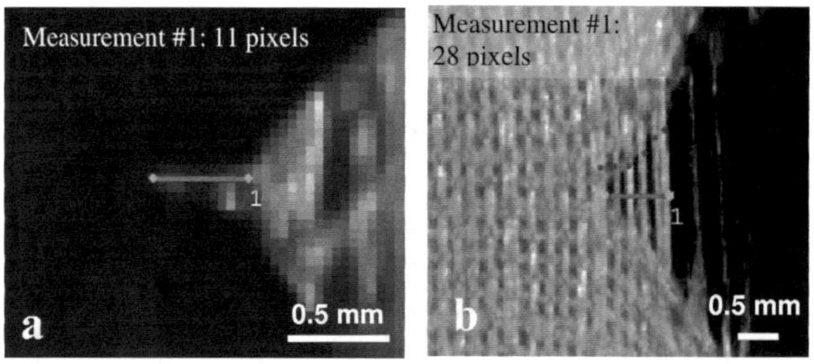

Figure 5. FASTCAM Viewer Software functions: (a) Measuring the pop-in crack (0.45 mm); (b) Counting the yarns (N=6) and crack wake bridging width (0.75 mm)

4. Conclusion

Tear strength is a fundamental characteristic used to assess the serviceability and quality of textiles. High-speed imaging analysis is a promising technique to observe and analyse the tearing behaviour of laminated fabrics, which is not possible during a test observed in real-time. This analysis allows for the examination of the tear strength phenomena on the fabric and membrane sides of laminated fabrics and helps visualize the behaviour of yarns before breaking, order of yarns breaking, pop-in crack, crack propagation, wrinkles, and thinning. Also, the FASTCAM Software helps quantify and measure many characteristics of the tearing behaviour such as the pop-in crack, crack wake bridging width, the number of yarns in the crack wake before breaking.

5. Acknowledgments

The authors gratefully acknowledge the financial support provided by the Natural Sciences and Engineering Research Council of Canada (NSERC) through its Discovery Program [Project Number: NSERC RGPIN-2019-05583], and the Canadian Foundation for Innovation (CFI) through the John R. Evans Leaders Fund [Project Number: 31331].

References

[1] Smith W C 2010 Overview of textile coating and lamination. In W. C. Smith (ed.), *Smart Textile Coatings and Laminates* (2nd ed.). Woodhead Publishing . Chapter 1 pp 3-9
[2] Painter C J 1996 Waterproof, Breathable Fabric Laminates: A Perspective from Film to Market Place. *Journal of Coated Fabrics*, **26**(2), pp 107–130.
[3] Scott R A 1995 Coated and laminated fabrics, in: Carr, C.M. (Ed.), *Chemistry of the Textiles Industry*. Springer Science & Business Media, pp 210–248.
[4] Lawson J R 1997 Fire fighters' protective clothing and thermal evironments of structural fire fighting. *Performance of protective clothing*, **6**
[5] Keighley J H 1985 Breathable Fabrics and Comfort in Clothing. *Journal of Coated Fabrics* **15** pp 89–104.
[6] Shaid A, Wang L, Padhye R 2018 Textiles for Firefighting Protective Clothing, in: Song, G., Wang, F. (Eds.), *Firefighters' Clothing and Equipment: Performance, Protection, and Comfort*. Routledge, pp. 1–23.
[7] Triki E, Dolez P, Vu-Khanh T, 2011 Tear resistance of woven textiles – Criterion and mechanisms. *Composites Part B: Engineering* **42**, pp 1851–1859.
[8] Eltahan E 2018. Structural parameters affecting tear strength of the fabrics tents. *Alexandria Engineering Journal* **57**, pp 97–105.
[9] Eryuruk S H, Kalaoğlu F, 2015 The Effect Of Weave Construction On Tear Strength Of Woven Fabrics. *Autex Research Journal* **15**, pp 207–214.
[10] Chen J, Chen W, Zhou H, Zhao B, Wang M, Sun W, He S, 2018 Fracture failure analysis and bias tearing strength criterion for a laminated fabric. *Journal of Industrial Textiles* **47**, pp 1496–1527.
[11] Bellinson H R, 1937 Simplified Trapezoid Tearing Test. *Textile Research* **7**, pp 208–210.
[12] Hamkins C P, Backer S, 1980 On the Mechanisms of Tearing in Woven Fabrics. *Textile Research Journal* **50**, pp 323–327.
[13] Hager O B, Gagliardi D D, Walker H B, 1947 Analysis of Tear Strength. *Textile Research Journal* **17**, pp 376–381.
[14] Turl L H, 1956 The Measurement of Tearing Strength of Textile Fabrics. *Textile Research* Journal **26**, pp 169–176.
[15] Steele R, Gruntfest I J, 1957 An Analysis of Tearing Failure. *Textile Research Journal* **27**, pp 307–313.
[16] Munevar-Ortiz L, Batcheller J, Dolez P I, 2021. Influence of Specimen Size on the Tear Strength of Fabrics by the Trapezoid Procedure. *Journal of testing and Evaluation* **50**, pp 1415-1424.
[17] Stenmark N, 2016. Failure Analysis of Plastics. *Stork Materials Technology*
[18] Lewis P R, 2016. Introduction, in: *Forensic Polymer Engineering*. Elsevier, pp 1–31.
[19] Standard Test Method for Tearing Strength of Fabrics by Trapezoid Procedure, ASTM D5587-15(2019) (West Conshohocken, PA: ASTM International).

SILICA-XEROGEL COATED COTTON AS MULTIFUNCTIONAL TEXTILE: DEVELOPMENT AND CHARACTERIZATION

M. ASSAL[1,2,*], M. EL WAZNA [1,2], K. OUMGHAR[2,3], M. EL BOUCHTI[2], A. EL BOUARI [1], O.CHERKAOUI [2]

[1] Laboratory of Physical Chemistry, Materials and Catalyst (LCPMC), Faculty of sciences Ben M'Sik - HASSAN II University of Casablanca, Casablanca, Morocco

[2] Laboratory textile materials research (REMTEX), Higher School of Textile and clothing Industries, Casablanca Morocco

[3]Laboratory of Physics of Condensed Matter (LPMC), Faculty of Sciences Ben M'Sik, Hassan II University, Casablanca, Morocco

Corresponding Author*: mohamed.assal-etu@etu.univh2c.ma

ABSTRACT:

Silica xerogel has attracted increasingly more attention in the field of thermal insulation as its extraordinary properties such as thermal conductivity, fire resistance, hydrophobicity and low density. This work focuses on the development of new insulating material, based on the xerogel as charge for cotton fabric. The composite was developed using vacuum molding process. The microstructure has been determined in each materials using SEM observations, and have a good hydrophobicity which the contact angle is 125°. The results obtained from this study can be useful to develop new low cost, sustainable, light product, insulation and environmentally friendly materials.

Keywords: Cotton fabric, Hydrophobicity, Physical properties, Porosity, Silica xerogel.

1. INTRODUCTION:

A deal of research [1-5] has been devoted to innovative materials with excellent thermal insulation and fire resistance. Xerogels are considered among the most promising high-performance products for various applications due to their low densities [6], hydrophobicity [7], low thermal conductivity [4]. However, their low mechanical properties have limited their use [5].

Xerogels are defined as a microporous backbone, assimilated to a two phase system (solid and gas) [8]. Prakash C. Thapliyal [9] reported that these materials are gels in which the liquid has been replaced by air, with very moderate shrinkage of a network solid. The combination of structural characteristics gives Xerogels special properties such as filtration and thermal insulation [10]. Deniz Bozoglu and al[11] reported that PU composite film with 7% aerogel had the highest contact angle (105°).

2. MATERIALS AND METHODS:

2.1. Materials:

Cotton fabric, Tetraethyl orthosilicate (TEOS) (98%, M=208,33 g/mol), ethanol (99.8%, M=46,07 g/mol), ammonium hydroxide NH_4OH (95%, M=35,04 g/mol) and Hydrochloric acid (0.01M, M=36,458 g/mol). All chemicals were purchased from Sigma-Aldrich, they are analytically pure and used without any further purification.

2.2. Methods:

The preparation method used is sol-gel [12] detailed on next figure (figure1):

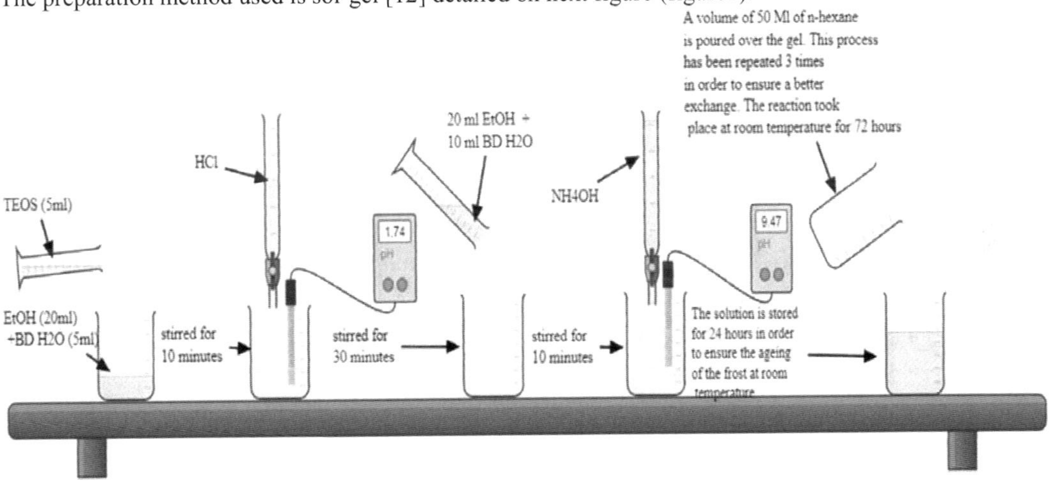

Figure 1: preparation method of silica-xerogel

Finally the gels were dried at a temperature of 60°C for 48 hours. The samples were developed using a coating process, which uses 10% of silica xerogel and 90% of poly acrylate.

3. RESULTS AND DISCUSSION:

3.1. Scanning Electron Microscopy (SEM) with Energy Dispersive X-Ray Analysis (EDX) :

Figure 2 show SEM images of silica xerogel obtained with secondary electron detection at accelerating voltages of 15 KV. Almost all the silica xerogel particles were spherical, and their size was about 10 nm. The pore structure of the xerogel, figure 3 show EDX-SEM results of prepared sample; confirm that the sample obtained contains of silice (Si) and oxygen (O) with an atomic percentage respectively of 23.79% and 76.21%.

Figure 2: SEM images of xerogel

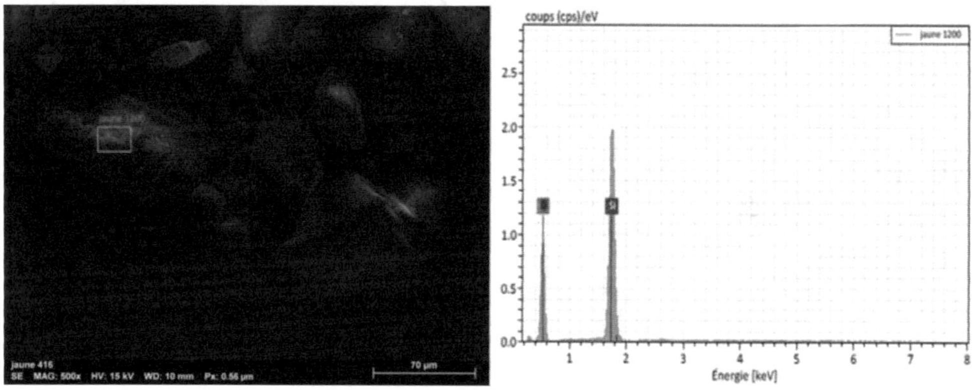

Figure 3: SEM EDX of silica xerogel

3.2. FT-IR Results:

FT-IR spectroscopy was used to study the functional groups present on the surface of hydrophobic silica xerogel. The functional groups present on the surface of the hydrophobic silica xerogel were studied. Figure 4 shows the FT-IR spectra of each synthesized silica xerogel. The absorption due to the C-H bond was observed at 1360 cm-1 , while the absorption due to the Si-C bond was observed at 1250 cm-1 and 839-757 cm-1. Similarly, the absorption corresponding to the Si-O bond was confirmed at 950 cm-1. On the other hand, the absorption features corresponding to the O-H bond at 3660 cm-1.

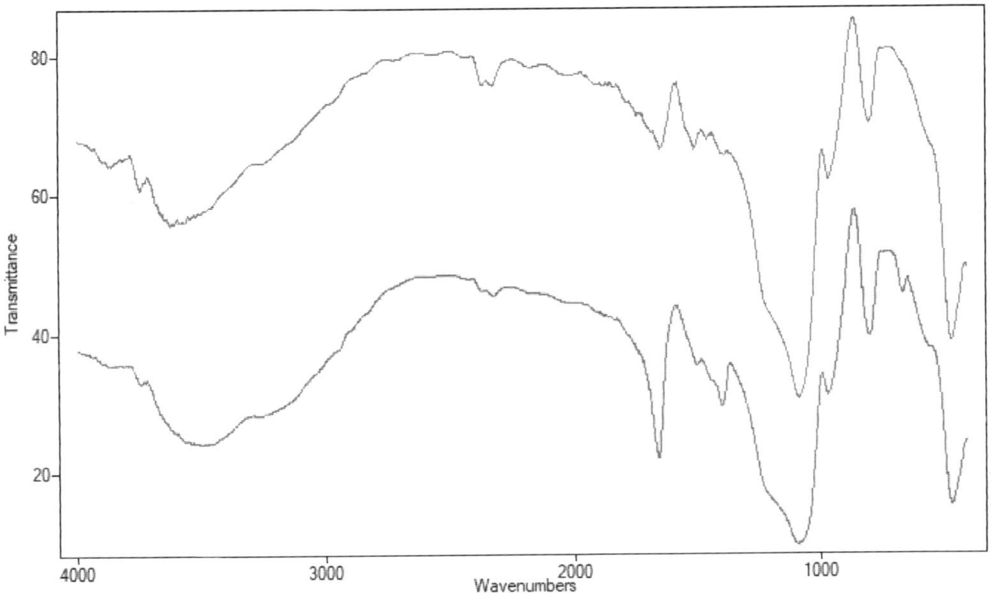

Figure 4: infrared results of silica xerogels

3.3. Contact angle results:

Figure 5 shows that after coating there is a great improvement in the contact angle with a shift from 45.5° for the cotton to 125.1° for treated cotton, with a difference superior to 75°. This shows the transition of the textile state from hydrophilic to hydrophobic.

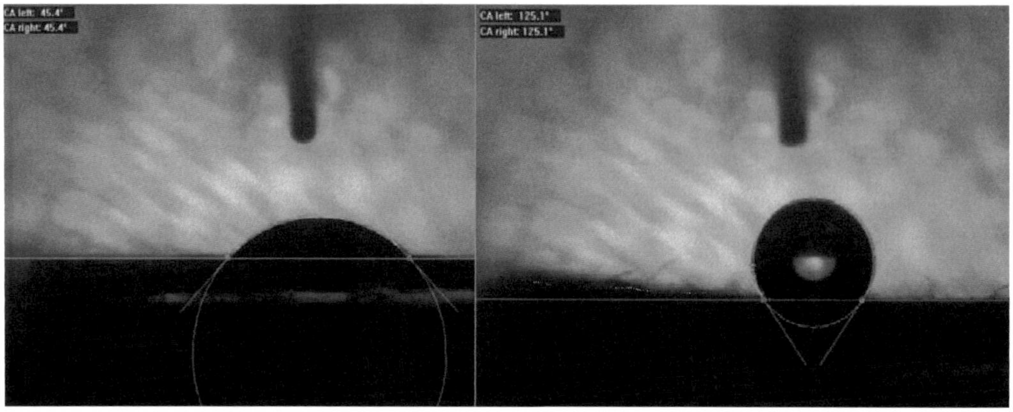

Figure 5: contact angle of composites

3.4: thermal conductivity measurement:

According to the thermal conductivity results there is a big difference between cotton and coated cotton due to the presence of xerogels, he value of thermal conductivity decreases from 0.398 for cotton to 0.136 for coated cotton, that make coated cotton more resistant thermally.

Table 1: value of thermal conductivity

References	cotton	Coated cotton
Thermal conductivity ($W.m^{-1}.k^{-1}$)	0.398	0.136
Thermal resistance ($m^2.k.w^{-1}$)	3.1	9.2

4. **CONCLUSION:**

Initial results show good adhesion between the sample components, which has been confirmed by the SEM plates. The exchange as well as the modification has been proven by FTIR spectroscopy, resulting in the hydrophobic behavior.

The results obtained from this study can be useful to develop new low cost, sustainable, light product, insulation and environmentally friendly materials.

REFERENCES :

[1] M. S. Al-Homoud, "Performance characteristics and practical applications of common building thermal insulation materials," Building and environment, vol. 40, no. 3, pp. 353-366, 2005.
[2] I. Cetiner and A. D. Shea, "Wood waste as an alternative thermal insulation for buildings, "Energy and Buildings, vol. 168, pp. 374-384, 2018.
[3] Z. L. Yu et al., "Fire-retardant and thermally insulating phenolic-silica aerogels," Angewandte Chemie International Edition, vol. 57, no. 17, pp. 4538-4542, 2018
[4] G. Jia, Z. Li, P. Liu, and Q. Jing, "Preparation and characterization of aerogel/expanded perlite composite as building thermal insulation material," Journal of Non-Crystalline Solids, vol. 482, pp. 192-202, 2018.
[5] U. Berardi, "Aerogel-enhanced systems for building energy retrofits : Insights from a case study," Energy and Buildings, vol. 159, pp. 370-381, 2018.
[6] S. Liu, K. Zhu, S. Cui, X. Shen, and G. Tan, "A novel building material with low thermal conductivity : Rapid synthesis of foam concrete reinforced silica aerogel and energy performance simulation," Energy and Buildings, vol. 177, pp. 385-393, 2018.
[7] K. Nocentini, P. Biwole, and P. Achard, "Silica Aerogel Blankets as Super insulating Material for Developing Energy Efficient Buildings," in Sustainability through Energy-Efficient Buildings: CRC Press, 2018, pp. 151-164.
[8] J. P. Vareda, A. Lamy-Mendes, and L. Durães, "A reconsideration on the definition of the termaerogel based on current drying trends," Microporous and Mesoporous Materials, vol. 258,pp. 211-216, 2018.
[9] P. C. Thapliyal and K. Singh, "Aerogels as promising thermal insulating materials: An overview," Journal of materials, vol. 2014, 2014.
[10] M. A. Aegerter, N. Leventis, and M. M. Koebel, Aerogels handbook. Springer Science & Business Media, 2011.
[11] Structural and dielectrical characterization of low-k polyurethane composite films with silica aerogel, Deniz Bozoglu, Huseyin Deligoz, Kemal Ulutas, SahinYakut, Deniz Deger. Journal of Physics and Chemistry of SolidsVolume 130, July 2019, Pages 46-57
[12] Aerogel-like silica powders by combustion of sol-gel derived alcogels. Martin Timuska, Triin Kangura, Janis Locsb, Andris Šutkac, Martin Järvekülga Microporous and Mesoporous Materials Volume 315, February 2021, 110895
[13] In situ one-step construction of monolithic silica aerogel-based composite phase change materials for thermal protection
Panpan Liu, Hongyi Gao, Xiao Chen, Dangjia Chen, Junjun Lv, Mengyi Han, Piao Cheng, Ge Wang. Composites Part B: Engineering Volume 195, 15 August 2020, 108072

AUTHOR INDEX